NIKOLA TESLA

My Inventions and
Other Writings

Introduction by SAMANTHA HUNT

PENGUIN BOOKS

PENGUIN BOOKS
Published by the Penguin Group
Penguin Group (USA) Inc., 375 Hudson Street,
New York, New York 10014, U.S.A.
Penguin Group (Canada), 90 Eglinton Avenue East, Suite 700, Toronto,
Ontario, Canada M4P 2Y3 (a division of Pearson Penguin Canada Inc.)
Penguin Books Ltd, 80 Strand, London WC2R oRL, England
Penguin Ireland, 25 St Stephen's Green, Dublin 2, Ireland (a division of Penguin Books Ltd)
Penguin Group (Australia), 250 Camberwell Road, Camberwell,
Victoria 3124, Australia (a division of Pearson Australia Group Pty Ltd)
Penguin Books India Pvt Ltd, 11 Community Centre, Panchsheel Park, New Delhi - 110 017, India
Penguin Group (NZ), 67 Apollo Drive, Rosedale, Auckland 0632,
New Zealand (a division of Pearson New Zealand Ltd)
Penguin Books (South Africa) (Pty) Ltd, 24 Sturdee Avenue,
Rosebank, Johannesburg 2196, South Africa

Penguin Books Ltd, Registered Offices:
80 Strand, London WC2R oRL, England

First published in Penguin Books 2011

10

Introduction copyright © Samantha Hunt, 2011
All rights reserved

The original publishers of the selections in this book are cited on page xxi.

LIBRARY OF CONGRESS CATALOGING IN PUBLICATION DATA
Tesla, Nikola, 1856–1943.
My inventions and other writings / Nikola Tesla ; introduction by Samantha Hunt.
p.   cm.
ISBN 978-0-14-310661-6
1. Tesla, Nikola, 1856–1943.   2. Electrical engineers—United States—Biography.   3. Inventors—
United States—Biography.   I. Hunt, Samantha.   II. Title.
TK140.T4A35 2012
621.3092—dc23
[B]   2011040456

Printed in the United States of America
Set in Sabon

Contents

Introduction

One evening in 1865 there was a buzzing in Smiljan, Croatia, a yearly plague of June beetles. The buzz kept the nine-year-old Nikola Tesla awake. In the morning he filled a glass jar to brimming with the big bugs and, selecting the hardest workers among them, glued the beetles' thoraxes to the gear of a small motor he'd built. The motor turned and Tesla's heart raced, as he understood that he and his latest invention would usher in an era when humans could abandon all manual labor. The world would now be powered by indentured insects. Lost in his glorious scheme, Tesla did not notice another boy enter his workshop until the boy, out of jealousy or perhaps simple hunger, dipped his hand into the glass storage jar, pulled out a fistful of bugs and enjoyed a leggy snack. Nikola vomited in the bushes, and insects remain an uncolonized species to this day. But the idea of harnessing natural energy and a penchant for crafting nearly magical devices would always stay with the young inventor.

Set back from the intersection of Thirty-fourth Street and Eighth Avenue in New York City there's a plaque, lonely in the gathering noise, easily escaping notice. It is a wallflower placed so awkwardly high off the street that it is difficult to read.

"HERE DIED," the plaque begins, "THE GREAT YUGOSLAV-AMERICAN, SCIENTIST-INVENTOR, NIKOLA TESLA." Those two hyphens had to be cast in bronze in order to carry all that they leave unsaid. Still, they are a place to begin remembering this most forgotten man.

Quickly, Nikola Tesla is responsible for our modern electrical system—alternating current. He was an inventor, an engineer, a scientist and oddball. It was Tesla, not Marconi, who invented radio. Though history books tend to forget it, the U.S. Supreme Court, in a gavel drop that no one heard, upheld Tesla's patent a few months after his death. Yet nearly everyone passing below this quiet plaque has no idea who Tesla was. All the Nikola Tesla High Schools that should pepper America are still figments. In truth, fairly few things bear his name. There is the international unit of magnetic flux density, the tesla, most often used by MRI technicians. It is a nice tribute but it is no watt, no volt.

Great masses of people pass below the plaque never knowing that Tesla lived most of his life alone in New York City hotel rooms. They don't know that he believed he was married to a pigeon or that he was friends with Mark Twain. And they don't know that, more than any one man, Nikola Tesla is responsible for the twentieth century.

When I first encountered *My Inventions* it was as an implausible work titled *The Strange Life of Nikola Tesla*. I dismissed the text as an invention itself, one concocted by a flamboyantly imaginative fan of Tesla's—a fairly common species. Sentences like "When I drop little squares of paper in a dish filled with liquid, I always sense a peculiar and awful taste in my mouth" convinced me that *The Strange Life of Nikola Tesla* was some sort of hoax. The story it told was too weird to be his. An engineering genius could never have drafted such a magnificently unscientific text—one that sometimes reads as if it has been written by a carnival barker. "And now I will tell of one of my feats with this antique implement of war which will strain to the utmost the credulity of the reader." Indeed.

But *The Strange Life of Nikola Tesla* was not a fabrication. Though that title had been added after his death, the text was in fact Tesla's work, first published serially in 1919 in *The Electrical Experimenter* magazine under the title "My Inventions." These essays tell the story of his early life, the rotary magnetic field, the Tesla coil and transformer. Each installment is a wondrous hybrid: part autobiography, part science, part *ars poetica* filled with earnest confessions and self-examinations frank as a child's. Stories of his boyhood cunning in catching rats, dueling with cornstalks, or attempting to fly off a barn roof mingle with sentences like "It is a resonant transformer with a secondary in which the parts, charged to a high potential, are of considerable area and arranged in space along ideal enveloping surfaces of very large radii of curvature, and at proper distances from one another thereby insuring a small electric surface density everywhere so that no leak can occur even if the conductor is bare." One paragraph alone traipses through clockworks, guns, and Serbian poetry. Tesla was a true liberal artist; his intelligence was unspecialized; his genius general. Voltaire exists alongside voltage. Engineering is poetry, as Tesla would broker no separation between art and science.

Perhaps because of this his writing can read like a work of science fiction:

> I may mention that only recently an odd looking gentleman called on me with the object of enlisting my services in the construction of world transmitters in some distant land. "We have no money," he said, "but carloads of solid gold and we will give you a liberal amount." I told him that I wanted to see first what will be done with my inventions in America, and this ended the interview. But I am satisfied that some dark forces are at work.

The story he makes of his life and ideas is so engaging that it is hard not to ask, Is this real? And then, Does it matter? Tesla was, after all, an inventor, and we would never deign to question the reality of one of his patents.

When originally published in the pages of *The Electrical Experimenter,* "My Inventions" mingled with an astonishing variety of colorful articles: "Soldiers' Ills Cured by Electricity," "Will Man Freeze the Earth to Death?", "Women Now Trained as Meter Maids," "Wood Finishing for the Amateur," and "Home Experiments in Radio-Activity." One *Electrical Experimenter* editorial states, "Conditions on Mars we know by direct observation as well as deduction are favorable for life, and we may be certain that it exists there." Advertisements urge the reader to "Humanize Your Talking Machine" or find the cure for all ills with Violet-Rays. There's an ad for a nose corset alongside one where Lionel Strongfort and his fitness regimen counsel, "Don't commit a crime against the woman you love." Curiosity, self-reliance, and naïveté were alive and well in 1919. "My Inventions" and the other essays included in this volume, like the back pages of a comic book, are full of the incredulous. Wonder abounds. No cynics allowed.

Tesla's world is long gone. The distance between then and now creates a gorgeous atmospheric haze. When he presents the past's notion of the future, it is hard not to mark how near or far off his predictions landed. There is a sense of nostalgia for a time that has not yet materialized and maybe never will: the future as Tesla imagined it, a place where machines save us all, meals are taken as efficient tiny pills, power and energy travel wirelessly around the globe, and world peace is a given.

The Electrical Experimenter, a technical science monthly that would later be absorbed into *Popular Mechanics,* was edited by a man named Hugo Gernsback, who felt that any "real electrical experimenter, worthy of the name" would, of course, be devising a plan for the future. And so Tesla was. Always. Since the only limits in his laboratory were financial, he was happy to entertain the possible, even when ridiculed. Can we speak to beings in outer space? Can we photograph thought or build a stationary ring around the equator? Why don't we try?

When measured against the cynicism of the twenty-first century and the corporate, governmental, and academic controls now placed on science, Tesla's open mind and laboratory seem like rarities. Consider the number of painters and writers, playwrights and performance artists all working on masterpieces today, all of them dealing with mystery and possibility. But where is the young scientist among them who toils in her garage, trying to mix her DNA with a great blue heron's? Somewhere between Tesla and now, invention has lost its illicitness. Somewhere art and science have parted ways, leaving the world to wonder where Tesla's descendants, all the poet-inventors, are hiding.

In his youth Tesla studied with a staggering appetite, memorizing *Faust* while compulsively observing the physical environment around him. His father, concerned about his son's health, forbade him the use of candles so that he'd not be able to read. Tesla fashioned his own candles and kept on reading. He studied himself into a number of illnesses and nervous conditions, including an obsession with germs and the number three. In the latter half of the nineteenth century, there was so much that needed inventing. What pressure Tesla must have felt as a young man, convinced that he could fashion the most efficient motor for an electrical generator, certain he could build a tower that would wirelessly transmit not only information but power. His passion was a danger to his health. While hoping to solve the problem of alternating current, he left himself open and entirely vulnerable.

Having always enjoyed highly keen senses, in 1881 Tesla was deluged. He writes that he "could hear the ticking of a watch with three rooms between me and the time-piece. A fly alighting on a table in the room would cause a dull thud in my ear. A carriage passing at a distance of a few miles fairly shook my whole body." These waves of raw sensitivity proved to be the labor pains of our modern electrical system. In the throes of this condition, he thought of a solution. He saw his alternating current motor whirling in the air before his eyes and was certain, without even building it, that it would work.

Tesla soon left for the United States with dreams of fabricating his motor. His journey, as he tells it, was a series of unbelievable events. Though he was robbed while traveling to the port, he did not turn back. He arrived on American shores with four centimes in his pocket. Strolling from Castle Clinton, New York's early immigration center, north to Edison's laboratory, he encountered a man cursing a broken machine on the street. Tesla swiftly fixed the gadget and the man paid him twenty dollars, an implausible sum for 1884. When Tesla arrived at Edison's lab, he was immediately hired by the great inventor—that very day—repairing dynamos and increasing efficiency in the leaky lab. Edison promised to pay Tesla fifty thousand dollars once the job was done. Tesla toiled for months, arriving at ten in the morning and working until five the following morning, going without sleep. When he was finished, Tesla went to collect his pay. Edison began to laugh, claiming that Tesla did not understand the American sense of humor. He refused to pay the fifty thousand dollars. Tesla's dream inventions crashed to the already cluttered floor. He resigned and spent the following year digging ditches, the AC motor spinning in his thoughts all the while. It was a time of such darkness that the inventor rarely spoke of it later in life.

More than determined, he slowly climbed out of the ditch where Edison had left him. Tesla cobbled together space, money, and investors. While he began to see his visions made manifest, his relationship with money would continue to be contentious, as Tesla rarely protected his patents. Like some proto–open source advocate, he believed his inventions belonged to the world, not just him. It is rumored that after Marconi sent the first wireless letter S across the ocean, an engineer working for Tesla chided him, saying, "Looks like Marconi got the jump on you." Tesla's answer was surprising. "Marconi is a good fellow. Let him continue. He is using seventeen of my patents." The well-connected Marconi would go on to win the Nobel Prize for radio, despite the solid fact that Tesla invented it. And history gets written in the strangest of ways.

Meanwhile, Edison was trying his best to discredit AC. With a campaign of fear, Edison aimed to squash the burgeoning technology because he believed that his light bulbs would not work on it. The War of the Currents roared. Edison electrocuted the animals of Menlo Park and built the first electric chair to be used at Sing-Sing prison in order to demonstrate AC's dangerous properties. Instead Edison's machine demonstrated only incompetence. The unfortunate William Kemmler, sentenced to die for murdering Tillie Ziegler, suffered an extended half death until nearly an hour after the process had begun.

The great test of Tesla's alternating current motor came at the 1893 Chicago World's Columbian Exposition. Edison entered a bid to light the fair with DC power. Tesla and George Westinghouse (Tesla had sold his patents to Westinghouse) entered a bid for AC. The fair would be the first time many Americans would experience the delights of electricity firsthand. AC, able to perform far more efficiently at a far cheaper price, won the bid handily so that when President Grover Cleveland touched a golden lever, firing up Tesla's dynamos, turning on more than 200,000 light bulbs in the White City, America took notice of the gangly Serbian. He and Westinghouse were awarded the contract to harness Niagara Falls and, for better or worse, the age of electricity began.

The influential occultist Aleister Crowley asked us to "Please remember that science is majick." The trouble Tesla would have with Crowley's statement was why call it magic at all? Why not give the wonders of the world their due by labeling them "science"? Why do we not simply believe that more is possible? With his inventions often arriving in explosive bursts of vision, it is no surprise that Tesla had an interest in the psychical. In his 1900 essay "The Problem of Increasing Human Energy" (included in this volume) he writes:

> We are all one. Metaphysical proofs are, however, not the only ones which we are able to bring forth in support of this idea. Science, too, recognizes this connectedness of separate individuals, though not quite in the same sense as it admits that the suns, planets, and moons of a constellation are one body, and there can be no doubt that it will be experimentally confirmed in times to come, when our means and methods for investigating psychical and other states and phenomena shall have been brought to great perfection.

This furthers an idea from his lecture "The Action of the Eye" (also included in this volume), where he contemplates mind reading in terms both scientific and poetic. Though he makes some egregious statements about eye luminosity equaling creativity, his question stands: if thought is electrical, why can't we record it and measure it?

From all that seems beautiful and strange in Tesla's writings, there is a theme that emerges: science versus the supernatural. Extraordinary things happened to Tesla because he had extraordinary powers of observation. The supernatural is super, yes, but more important, it is natural.

Tesla's open mind sometimes lands him in trouble as his

legacy often gets claimed by a lunatic fringe. This does little to aid his assimilation into history books. One biography, queerly printed in Kelly green ink, asserts, "Nikola Tesla was not an Earth man. The space people have stated that a male child was born on board a space ship which was on a flight from Venus to the Earth in July, 1856. The little boy was called Nikola."

Poor Tesla. I suppose it is a compliment to believe he is so far above us that he comes from Venus, but why not instead insist that such brilliance is human?

While Tesla might have been the darling of the Chicago fair, the new century was not kind to him. Westinghouse, facing pressure from J. P. Morgan, had asked Tesla to tear up their contract, the one that gave Tesla a percentage of every horsepower of AC-generated electricity ever sold. Westinghouse claimed that alternating current would not survive the crucible of Morgan's capitalism with Tesla's contract in place. So Tesla, in order to see his invention live, tore up the contract.

He spent the last ten years of his life at the Hotel New Yorker. When it opened in 1930 it was the tallest building in New York City, a monument to the ambition and decadence of the Jazz Age. At forty-three stories high, it had its own power generator. The kitchen was an entire acre. There were five restaurants, ten private dining rooms, two ballrooms, and an indoor ice skating rink. Conveyor belts whisked dirty dishes through secret passageways down to fully automated dishwashers. Four stories below ground, bed sheets and tablecloths were miraculously laundered, dried, ironed, and folded without ever touching a human hand. Everything about the hotel was efficient, futuristic. It was perfect for Tesla—except that by the time he arrived at the New Yorker in 1933, he was destitute.

I check myself into the hotel. I request his room, 3327, and am twice surprised. First, little marks the room as his and, second, the chamber is quite modest. I take a bath in what was once his tub, trying to soak up his materiality. Nikola Tesla sat alone in this room for ten years. He opened this door, breathed this air, saw this view. He walked all over the island of Manhattan. The hotel hums with power and yet I don't find him there. Instead, I find the same question. Why has he been so forgotten?

Perhaps Tesla's ideas were too terrific, too far before his time. He tinkered with a number of dreamily ingenious schemes, some realized, some still dreams: to control objects remotely, light the oceans, photograph thoughts, communicate with life in outer space, harvest free energy from the Earth's atmosphere, control the weather with electricity, build a ring around the equator that, by remaining stationary while the planet rotates, would make it possible to travel around the entire world in one day. He was an environmentalist in the age of robber barons. "It is our duty to coming generations to leave this store of energy intact for them, or at least not touch it until we have perfected processes for burning coal more efficiently." ("The Problem of Increasing Human Energy") Tesla, who died two years before Hiroshima and Nagasaki, spoke frequently of his plans to build a death ray that, in presenting the threat of our total annihilation, would guarantee world peace. Innocently he writes, "Not even the most war-crazed power ever would venture such an immeasurable and unpardonable outrage against mankind, involving as it would the burning of helpless women and children and noncombatants." ("Tesla Would Pour Lightning . . .")

He goes on to write that rather than holding peace through the stasis of fear, his wireless technology would expose our undeniable connections and assure a real peace.

> If we were to release the energy of atoms or discover some other way of developing cheap and unlimited power at any point of the globe this accomplishment, instead of being a blessing, might bring disaster to mankind in giving rise to dissension and anarchy which would ultimately result in the enthronement of the hated regime of force. The greatest good will comes from technical improvements tending to unification and harmony, and my wireless transmitter is preeminently such. By its means the human voice and likeness will be reproduced everywhere.

Xenophobia certainly plays a role in Tesla's having been forgotten in America. Edison once, unable to locate Smiljan on a map, asked him in all sincerity if he had ever eaten human flesh. It was difficult for a vanilla American public, reeling from war, to lionize a foreign eccentric, an eternal bachelor whose best friends were the pigeons of Bryant Park. Edison, the master marketer, released "The Edison Polka," a tune commissioned to sell his phonographs. He gave the people something to dance to while Tesla, with talk of death rays, lightning bolts, and extra-terrestrials, gave a war-weary nation the creeps.

And so his feelings toward humans are understandably complicated. Though he was happy to house the sickliest of New York's pigeons, he couldn't bear the touch of human hair or the sight of a woman wearing pearl earrings. He believed that inventors, to stay true to their calling, should never marry. His life was marked by a series of episodes where trust was betrayed. It is no wonder he retreated to the solitude of his hotel room, studying humankind from afar.

He had few friends and was often torn, feeling both wonder and disgust toward other people. In "The Problem of Increasing Human Energy" he reduces humans to the phrase "units of higher value." Tesla is a paradox. While guilty of writing, "When I am all but used up I simply do as the darkies, who 'naturally fall asleep while white folks worry,'" he also states: "Peace can only come as a natural consequence of universal enlightenment and merging of races, and we are still far from this blissful realization." For all his love of humanity, it seems Tesla did not much care for humans.

In 1915, a *New York Times* reporter phoned Tesla to inform him that he and Edison were to share that year's Nobel Prize. Tesla was concerned about splitting the award with his rival, but he so clearly deserved the prize that his initial fears never could have presaged what was to come next. The report was erroneous. The 1915 Nobel Prize in Physics went to Sir William Henry Bragg and William Lawrence Bragg "for their services in the analysis of crystal structure by means of X-rays." Tesla was never recognized by the Nobel committee.

I am unwilling to accord to some smallminded and jealous individuals the satisfaction of having thwarted my efforts. These men are to me nothing more than microbes of a nasty disease. My project was retarded by laws of nature. The world was not prepared for it. It was too far ahead of time. But the same laws will prevail in the end and make it a triumphal success.

When he was a boy, Tesla climbed to the top of a barn and, holding on to nothing more than an umbrella, jumped off. Did he fly? No. But he demonstrated an epic fearlessness in the face of failure. Tesla tried the impossible. Sometimes he failed. Sometimes he didn't.

Darkness makes me think of Tesla, particularly that specific pain experienced when a blackout brings our rolling, connected world to a disconnected halt.

In Tesla's honor I try to write this in the dark, working in rooms without lights or heat, without a computer. Every misstep I make casts me back further in time, closer to Tesla and the world he experienced as a young man. If I sit like this long enough, I'll start to forget what I now know too well so that next time I'm well lit and warm, all marvel and magnificence will be restored to the technology I take for granted. I'll be flush with awe again each time my fingertip brushes a light switch. Perhaps then I'll be able to conjure a dark New York City, 1884, the day he stepped off the SS *Saturnia*.

Eventually I have to turn on the light. This simple gesture doesn't quite bring the flood of renewed disbelief and awe I'd hoped. But it does have me wondering again what electricity is exactly. Not even Tesla could say. "The day when we shall know exactly what 'electricity' is will chronicle an event probably greater and more important than any other recorded in the history of human race." ("The Action of the Eye")

My house is filled with tiny lights. At night red, green, and white flashes glow from the modem, the cell phone charger, the wireless baby monitor, the computer, coffeemaker, DVD player, even the switch on the surge protector. And there is a general hum, a sound I've grown so accustomed to. It is the sound of a house alive and breathing.

Outside my window there is a conjunction of power lines. I

follow the route of these wires as far as my eye can. I close my eyes and follow it even farther, making a road back to him, wirelessly. The road passes through the Hotel New Yorker and the Waldorf. It passes through the *Saturnia,* which took weeks to get to America. Men in suits, men in hats. Soon the road is so long it becomes a black and white road, a road before color. A road without tarmac or cars or power lines that goes all the way back to the tiny village of Smiljan, Croatia, where a Serbian boy was born at the stroke of midnight to a minister and his wife. There on the road is a tall and handsome man: Nikola Tesla, inventor. He looks up, surprised after so many years to be recognized.

> Only thru annihilation of distance in every respect, as the conveyance of intelligence, transport of passengers and supplies and transmission of energy will conditions be brought about some day, insuring permanency of friendly relations. What we now want most is closer contact and better understanding between individuals and communities all over the earth.

Perhaps the best way to restore this great man to Earth, to us, is to read the world as he wrote it. Even if that world is still a long way off.

SAMANTHA HUNT

A Note on the Text

This edition of *My Inventions* reprints the original, authoritative text, as first serialized in The *Electrical Experimenter* magazine. Part I: "My Early Life" was first serialized in the February 1919 (volume VI, number 10) issue; Part II: "My First Efforts in Inventions" was first serialized in the March 1919 (volume VI, number 11) issue; Part III: "My Later Endeavors: The Discovery of the Rotating Magnetic Field" was first serialized in the April 1919 (volume VI, number 12) issue; Part IV: "The Discovery of the Tesla Coil and Transformer" was first serialized in the May 1919 (volume VII, number 1) issue; Part V: "The Magnifying Transmitter" was first serialized in the June 1919 (volume VII, number 2) issue; and Part VI: "The Art of Telautomatics" was first serialized in the October 1919 (volume VII, number 6) issue of The *Electrical Experimenter*.

The additional articles included in this edition of *My Inventions and Other Writings* are available at the New York Public Library. "Tesla Would Pour Lightning from Airships to Consume Foe" first appeared in *The Washington Post* on February 13, 1916. "The Action of the Eye" first appeared in the *American Journal of Photography* on September 1, 1893 (and was an extract from his paper "Action of the Eye," read before the Franklin Institute). "The Problem of Increasing Human Energy" first appeared in *Century Illustrated Magazine* in June 1900. This edition does not reprint the diagrams and photographs included in that original publication. Interested readers may access the original photographs and other images from the New York Public Library.

My Inventions

PART I

MY EARLY LIFE

The progressive development of man is vitally dependent on invention. It is the most important product of his creative brain. Its ultimate purpose is the complete mastery of mind over the material world, the harnessing of the forces of nature to human needs. This is the difficult task of the inventor who is often misunderstood and unrewarded. But he finds ample compensation in the pleasing exercises of his powers and in the knowledge of being one of that exceptionally privileged class without whom the race would have long ago perished in the bitter struggle against pitiless elements.

Speaking for myself, I have already had more than my full measure of this exquisite enjoyment, so much that for many years my life was little short of continuous rapture. I am credited with being one of the hardest workers and perhaps I am, if thought is the equivalent of labor, for I have devoted to it almost all of my waking hours. But if work is interpreted to be a definite performance in a specified time according to a rigid rule, then I may be the worst of idlers. Every effort under compulsion demands a sacrifice of life-energy. I never paid such a price. On the contrary, I have thrived on my thoughts.

In attempting to give a connected and faithful account of my activities in this series of articles which will be presented with the assistance of the Editors of the ELECTRICAL EXPERIMENTER and are chiefly addrest to our young men readers, I must dwell, however reluctantly, on the impressions of my youth and the circumstances and events which have been instrumental in determining my career.

Our first endeavors are purely instinctive, promptings of an

imagination vivid and undisciplined. As we grow older reason
asserts itself and we become more and more systematic and
designing. But those early impulses, tho not immediately productive, are of the greatest moment and may shape our very
destinies. Indeed, I feel now that had I understood and culti-
vated instead of suppressing them, I would have added substantial value to my bequest to the world. But not until I had
attained manhood did I realize that I was an inventor.

This was due to a number of causes. In the first place I had
a brother who was gifted to an extraordinary degree—one of
those rare phenomena of mentality which biological investigation has failed to explain. His premature death left my parents
disconsolate. We owned a horse which had been presented to us
by a dear friend. It was a magnificent animal of Arabian breed,
possest of almost human intelligence, and was cared for and
petted by the whole family, having on one occasion saved my
father's life under remarkable circumstances. My father had
been called one winter night to perform an urgent duty and
while crossing the mountains, infested by wolves, the horse
became frightened and ran away, throwing him violently to
the ground. It arrived home bleeding and exhausted, but after
the alarm was sounded immediately dashed off again, returning to the spot, and before the searching party were far on the
way they were met by my father, who had recovered consciousness and remounted, not realizing that he had been lying in the
snow for several hours. This horse was responsible for my
brother's injuries from which he died. I witnest the tragic scene
and altho fifty-six years have elapsed since, my visual impression of it has lost none of its force. The recollection of his attainments made every effort of mine seem dull in comparison.

Anything I did that was creditable merely caused my parents
to feel their loss more keenly. So I grew up with little confidence in myself. But I was far from being considered a stupid
boy, if I am to judge from an incident of which I have still a
strong remembrance. One day the Aldermen were passing thru
a street where I was at play with other boys. The oldest of these
venerable gentlemen—a wealthy citizen—paused to give a sil-
ver piece to each of us. Coming to me he suddenly stopt and

commanded, "Look in my eyes." I met his gaze, my hand out-
stretched to receive the much valued coin, when, to my dismay,
he said, "No, not much, you can get nothing from me, you are
too smart." They used to tell a funny story about me. I had two
old aunts with wrinkled faces, one of them having two teeth
protruding like the tusks of an elephant which she buried in my
cheek every time she kist me. Nothing would scare me more
than the prospect of being hugged by these as affectionate as
unattractive relatives. It happened that while being carried in
my mother's arms they asked me who was the prettier of the
two. After examining their faces intently, I answered thought-
fully, pointing to one of them, "This here is not as ugly as the
other."

Then again, I was intended from my very birth for the
clerical profession and this thought constantly opprest me. I
longed to be an engineer but my father was inflexible. He was
the son of an officer who served in the army of the Great Napo-
leon and, in common with his brother, professor of mathemat-
ics in a prominent institution, had received a military education
but, singularly enough, later embraced the clergy in which
vocation he achieved eminence. He was a very erudite man, a
veritable natural philosopher, poet and writer and his sermons
were said to be as eloquent as those of Abraham a Sancta-Clara.
He had a prodigious memory and frequently recited at length
from works in several languages. He often remarked playfully
that if some of the classics were lost he could restore them. His
style of writing was much admired. He penned sentences short
and terse and was full of wit and satire. The humorous remarks
he made were always peculiar and characteristic. Just to illustrate, I may mention one or two instances. Among the help
there was a cross-eyed man called Mane, employed to do work
around the farm. He was chopping wood one day. As he swung
the axe my father, who stood nearby and felt very uncomfort-
able, cautioned him, "For God's sake, Mane, do not strike at
what you are looking but at what you intend to hit." On
another occasion he was taking out for a drive a friend who
carelessly permitted his costly fur coat to rub on the carriage
wheel. My father reminded him of it saying, "Pull in your

coat, you are ruining my tire." He had the odd habit of talking to himself and would often carry on an animated conversation and indulge in heated argument, changing the tone of his voice. A casual listener might have sworn that several people were in the room.

Altho I must trace to my mother's influence whatever inventiveness I possess, the training he gave me must have been helpful. It comprised all sorts of exercises—as, guessing one another's thoughts, discovering the defects of some form or expression, repeating long sentences or performing mental calculations. These daily lessons were intended to strengthen memory and reason and especially to develop the critical sense, and were undoubtedly very beneficial.

My mother descended from one of the oldest families in the country and a line of inventors. Both her father and grandfather originated numerous implements for household, agricultural and other uses. She was a truly great woman, of rare skill, courage and fortitude, who had braved the storms of life and past thru many a trying experience. When she was sixteen a virulent pestilence swept the country. Her father was called away to administer the last sacraments to the dying and during his absence she went alone to the assistance of a neighboring family who were stricken by the dread disease. All of the members, five in number, succumbed in rapid succession. She bathed, clothed and laid out the bodies, decorating them with flowers according to the custom of the country and when her father returned he found everything ready for a Christian burial. My mother was an inventor of the first order and would, I believe, have achieved great things had she not been so remote from modern life and its multifold opportunities. She invented and constructed all kinds of tools and devices and wove the finest designs from thread which was spun by her. She even planted the seeds, raised the plants and separated the fibers herself. She worked indefatigably, from break of day till late at night, and most of the wearing apparel and furnishings of the home was the product of her hands. When she was past sixty, her fingers were still nimble enough *to tie three knots in an eyelash.*

There was another and still more important reason for my late awakening. In my boyhood I suffered from a peculiar affliction due to the appearance of images, often accompanied by strong flashes of light, which marred the sight of real objects and interfered with my thought and action. They were pictures of things and scenes which I had really seen, never of those I imagined. When a word was spoken to me the image of the object it designated would present itself vividly to my vision and sometimes I was quite unable to distinguish whether what I saw was tangible or not. This caused me great discomfort and anxiety. None of the students of psychology or physiology whom I have consulted could ever explain satisfactorily these phenomena. They seem to have been unique altho I was probably predisposed as I know that my brother experienced a similar trouble. The theory I have formulated is that the images were the result of a reflex action from the brain on the retina under great excitation. They certainly were not hallucinations such as are produced in diseased and anguished minds, for in other respects I was normal and composed. To give an idea of my distress, suppose that I had witnest a funeral or some such nerve-racking spectacle. Then, inevitably, in the stillness of night, a vivid picture of the scene would thrust itself *before* my eyes and persist despite all my efforts to banish it. Sometimes it would even remain fixt in space tho I pushed my hand thru it. If my explanation is correct, it should be possible to project on a screen the image of any object one conceives and make it visible. Such an advance would revolutionize all human relations. I am convinced that this wonder can and will be accomplished in time to come; I may add that I have devoted much thought to the solution of the problem.

To free myself of these tormenting appearances, I tried to concentrate my mind on something else I had seen, and in this way I would often obtain temporary relief; but in order to get it I had to conjure continuously new images. It was not long before I found that I had exhausted all of those at my command; my "reel" had run out, as it were, because I had seen little of the world—only objects in my home and the immediate surroundings. As I performed these mental operations for the second or third time, in order to chase the appearances from my vision, the remedy gradually lost all its force. Then I instinctively commenced to make excursions beyond the limits of the small world of which I had knowledge, and I saw new scenes. These were at first very blurred and indistinct, and would flit away when I tried to concentrate my attention upon them, but by and by I succeeded in fixing them; they gained in strength and distinctness and finally assumed the concreteness of real things. I soon discovered that my best comfort was attained if I simply went on in my vision farther and farther, getting new impressions all the time, and so I began to travel—of course, in my mind. Every night (and sometimes during the day), when alone, I would start on my journeys—see new places, cities and countries—live there, meet people and make friendships and acquaintances and, however unbelievable, it is a fact that they were just as dear to me as those in actual life and not a bit less intense in their manifestations.

This I did constantly until I was about seventeen when my thoughts turned seriously to invention. Then I observed to my delight that I could visualize with the greatest facility. I needed no models, drawings or experiments. I could picture them all as real in my mind. Thus I have been led unconsciously to evolve what I consider a new method of materializing invent- ive concepts and ideas, which is radically opposite to the purely experimental and is in my opinion ever so much more exped- itious and efficient. The moment one constructs a device to carry into practise a crude idea he finds himself unavoidably engrost with the details and defects of the apparatus. As he goes on improving and reconstructing, his force of concentra- tion diminishes and he loses sight of the great underlying prin- ciple. Results may be obtained but always at the sacrifice of quality.

My method is different. I do not rush into actual work. When I get an idea I start at once building it up in my imagination. I change the construction, make improvements and operate the device in my mind. It is absolutely immaterial to me whether I run my turbine in thought or test it in my shop. *I even note if it is out of balance.* There is no difference whatever, the results are the same. In this way I am able to rapidly develop and perfect a conception without touching anything. When I have gone so far as to embody in the invention every possible improvement I can think of and see no fault anywhere, I put into concrete form this final product of my brain. Invariably my device works as I conceived that it should, and the experiment comes out exactly as I planned it. In twenty years there has not been a single exception. Why should it be otherwise? Engineering, electrical and mechanical, is positive in results. There is scarcely a subject that cannot be mathematically treated and the effects calculated or the results determined beforehand from the available theoretical and practical data. The carrying out into practise of a crude idea as is being generally done is, I hold, nothing but a waste of energy, money and time.

My early affliction had, however, another compensation. The incessant mental exertion developed my powers of observation and enabled me to discover a truth of great importance. I had noted that the appearance of images was always preceded by actual vision of scenes under peculiar and generally very exceptional conditions and I was impelled on each occasion to locate the original impulse. After a while this effort grew to be almost automatic and I gained great facility in connecting cause and effect. Soon I became aware, to my surprise, that every thought I conceived was suggested by an external impression. Not only this but all my actions were prompted in a similar way. In the course of time it became perfectly evident to me that I was merely an *automaton* endowed with power of movement, responding to the stimuli of the sense organs and thinking and acting accordingly. The practical result of this was the art of *telautomatics* which has been so far carried out only in an imperfect manner. Its latent possibilities will, however, be eventually shown. I have been since years planning self-controlled automata and believe that mechanisms can be produced which will act as if possest of reason, to a limited degree, and will create a revolution in many commercial and industrial departments.

I was about twelve years old when I first succeeded in banishing an image from my vision by wilful effort, but I never had any control over the flashes of light to which I have referred. They were, perhaps, my strangest experience and inexplicable. They usually occurred when I found myself in a dangerous or distressing situation or when I was greatly exhilarated. In some instances I have seen all the air around me filled with tongues of living flame. Their intensity, instead of diminishing, increased with time and seemingly attained a maximum when I was about twenty-five years old. While in Paris, in 1883, a prominent French manufacturer sent me an invitation to a shooting expedition which I accepted. I had been long confined to the factory and the fresh air had a wonderfully invigorating effect on me. On my return to the city that night I felt a positive sensation that my brain had caught fire. I saw a light as tho a small sun was located in it and I past the whole night applying cold compressions to my tortured head. Finally the flashes diminished in frequency and force but it took more than three weeks before they wholly subsided. When a second invitation was extended to me my answer was an emphatic NO!

These luminous phenomena still manifest themselves from time to time, as when a new idea opening up possibilities strikes me, but they are no longer exciting, being of relatively small intensity. When I close my eyes I invariably observe first, a background of very dark and uniform blue, not unlike the sky on a clear but starless night. In a few seconds this field becomes animated with innumerable scintillating flakes of green, arranged in several layers and advancing towards me. Then there appears, to the right, a beautiful pattern of two systems of parallel and closely spaced lines, at right angles to one another, in all sorts of colors with yellow-green and gold predominating. Immediately thereafter the lines grow brighter and the whole is thickly sprinkled with dots of twinkling light. This picture moves slowly across the field of vision and in about ten seconds vanishes to the left, leaving behind a ground of rather unpleasant and inert grey which quickly gives way to a billowy sea of clouds, seemingly trying to mould themselves in living shapes. It is curious that I cannot project a form into this grey until the second phase is reached. Every time, before falling asleep, images of persons or objects flit before my view. When I see them I know that I am about to lose consciousness. If they are absent and refuse to come it means a sleepless night.

To what an extent imagination played a part in my early life I may illustrate by another odd experience. Like most children I was fond of jumping and developed an intense desire to support myself in the air. Occasionally a strong wind richly charged with oxygen blew from the mountains rendering my body as light as cork and then I would leap and float in space for a long time. It was a delightful sensation and my disappointment was keen when later I undeceived myself.

During that period I contracted many strange likes, dislikes and habits, some of which I can trace to external impressions while others are unaccountable. I had a violent aversion against the earrings of women but other ornaments, as bracelets, pleased me more or less according to design. The sight of a pearl would almost give me a fit but I was fascinated with the glitter of crystals or objects with sharp edges and plane surfaces. I would not touch the hair of other people except, perhaps, at the point of a revolver. I would get a fever by looking at a peach and if a piece of camphor was anywhere in the house it caused me the keenest discomfort. Even now I am not insensible to some of these upsetting impulses. When I drop little squares of paper in a dish filled with liquid, I always sense a peculiar and awful taste in my mouth. I counted the steps in my walks and calculated the cubical contents of soup plates, coffee cups and pieces of food,—otherwise my meal was unenjoyable. All repeated acts or operations I performed had to be divisible by three and if I mist I felt impelled to do it all over again, even if it took hours.

Up to the age of eight years, my character was weak and vacillating. I had neither courage or strength to form a firm resolve. My feelings came in waves and surges and vibrated unceasingly between extremes. My wishes were of consuming force and like the heads of the hydra, they multiplied. I was opprest by thoughts of pain in life and death and religious fear. I was swayed by superstitious belief and lived in constant dread of the spirit of evil, of ghosts and ogres and other unholy monsters of the dark. Then, all at once, there came a tremendous change which altered the course of my whole existence.

Of all things I liked books the best. My father had a large library and whenever I could manage I tried to satisfy my passion for reading. He did not permit it and would fly into a rage when he caught me in the act. He hid the candles when he found that I was reading in secret. He did not want me to spoil my eyes. But I obtained tallow, made the wicking and cast the sticks into tin forms, and every night I would bush the keyhole and the cracks and read, often till dawn, when all others slept and my mother started on her arduous daily task. On one occasion I came across a novel entitled "Abafi" (the Son of Aba), a Serbian translation of a well known Hungarian writer, Josika. This work somehow awakened my dormant powers of will and I began to practise self-control. At first my resolutions faded like snow in April, but in a little while I conquered my weakness and felt a pleasure I never knew before—that of doing as I willed. In the course of time this vigorous mental exercise became second nature. At the outset my wishes had to be subdued but gradually desire and will grew to be identical. After years of such discipline I gained so complete a mastery over myself that I toyed with passions which have meant destruction to some of the strongest men. At a certain age I contracted a mania for gambling which greatly worried my parents. To sit down to a game of cards was for me the quintessence of pleasure. My father led an exemplary life and could not excuse the senseless waste of time and money in which I indulged. I had a strong resolve but my philosophy was bad. I would say to him, "I can stop whenever I please but is it worth while to give up that which I would purchase with the joys of Paradise?" On frequent occasions he gave vent to his anger and contempt but my mother was different. She understood the character of men and knew that one's salvation could only be brought about thru his own efforts. One afternoon, I remember, when I had lost all my money and was craving for a game, she came to me with a roll of bills and said, "Go and enjoy yourself. The sooner you lose all we possess the better it will

be. I know that you will get over it." She was right. I conquered my passion then and there and only regretted that it
had not been a hundred times as strong. I not only vanquished
but tore it from my heart so as not to leave even a trace of desire.
Ever since that time I have been as indifferent to any form of
gambling as to picking teeth.

During another period I smoked excessively, threatening to
ruin my health. Then my will asserted itself and I not only
stopt but destroyed all inclination. Long ago I suftered from
heart trouble until I discovered that it was due to the innocent
cup of coffee I consumed every morning. I discontinued at once,
tho I confess it was not an easy task. In this way I checked and
bridled other habits and passions and have not only preserved
my life but derived an immense amount of satisfaction from
what most men would consider privation and sacrifice.

After finishing the studies at the Polytechnic Institute and
University I had a complete nervous breakdown and while
the malady lasted I observed many phenomena strange and
unbelievable.

PART II

MY FIRST EFFORTS
IN INVENTIONS

I shall dwell briefly on these extraordinary experiences, on account of their possible interest to students of psychology and physiology and also because this period of agony was of the greatest consequence on my mental development and subsequent labors. But it is indispensable to first relate the circumstances and conditions which preceded them and in which might be found their partial explanation.

From childhood I was compelled to concentrate attention upon myself. This caused me much suffering but, to my present view, it was a blessing in disguise for it has taught me to appreciate the inestimable value of introspection in the preservation of life, as well as a means of achievement. The pressure of occupation and the incessant stream of impressions pouring into our consciousness thru all the gateways of knowledge make modern existence hazardous in many ways. Most persons are so absorbed in the contemplation of the outside world that they are wholly oblivious to what is passing on within themselves. The premature death of millions is primarily traceable to this cause. Even among those who exercise care it is a common mistake to avoid imaginary, and ignore the real dangers. And what is true of an individual also applies, more or less, to a people as a whole. Witness, in illustration, the prohibition movement. A drastic, if not unconstitutional, measure is now being put thru in this country to prevent the consumption of alcohol and yet it is a positive fact that coffee, tea, tobacco, chewing gum and other stimulants, which are freely indulged in even at the tender age, are vastly more injurious to the national body, judging from the number of those who succumb. So, for instance, during my student years I gathered from the published necrologues in Vienna, the home of coffee drinkers, that deaths from heart trouble sometimes reached *sixty-seven per cent* of the total. Similar observations might probably be made in cities where the consumption of tea is excessive. These delicious beverages super-excite and gradually exhaust the fine fibers of the brain. They also interefere seriously with arterial circulation and should be enjoyed all the more sparingly as their deleterious effects are slow and imperceptible. Tobacco, on the other hand, is conducive to easy and pleasant thinking and detracts from the intensity and concentration necessary to all original and vigorous effort of the intellect. Chewing gum is helpful for a short while but soon drains the glandular system and inflicts irreparable damage, not to speak of the revulsion it creates. Alcohol in small quantities is an excellent tonic, but is toxic in its action when absorbed in larger amounts, quite immaterial as to whether it is taken in as whiskey or produced in the stomach from sugar. But it should not be overlooked that all these are great eliminators assisting Nature, as they do, in upholding her stern but just law of the survival of the fittest. Eager reformers should also be mindful of the eternal perversity of mankind which makes the indifferent *"laissez-faire"* by far preferable to enforced restraint. The truth about this is that we need stimulants to do our best work under present living conditions, and that we must exercise moderation and control our appetites and inclinations in every direction. That is what I have been doing for many years, in this way maintaining myself young in body and mind. Abstinence was not always to my liking but I find ample reward in the agreeable experiences I am now making. Just in the hope of converting some to my precepts and convictions I will recall one or two.

A short time ago I was returning to my hotel. It was a bitter cold night, the ground slippery, and no taxi to be had. Half a block behind me followed another man, evidently as anxious as myself to get under cover. Suddenly my legs went up in the air. In the same instant there was a flash in my brain, the nerves responded, the muscles contracted, I swung thru 180 degrees and landed on my hands. I resumed my walk as tho nothing had happened when the stranger caught up with me. "How old are you?" he asked, surveying me critically. "Oh, about fifty-nine," I replied. "What of it?" "Well," said he, "I have seen a cat do this but never a man." About a month since I wanted to order new eyeglasses and went to an oculist who put me thru the usual tests. He lookt at me increduously as I read off with ease the smallest print at considerable distance. But when I told him that I was past sixty he gasped in astonishment. Friends of mine often remark that my suits fit me like gloves but they do not know that all my clothing is made to measurements which were taken nearly 35 years ago and never changed. During this same period my weight has not varied one pound.

In this connection I may tell a funny story. One evening, in the winter of 1885, Mr. Edison, Edward H. Johnson, the President of the Edison Illuminating Company, Mr. Batchellor, Manager of the works, and myself entered a little place opposite 65 Fifth Avenue where the offices of the company were located. Someone suggested guessing weights and I was induced to step on a scale. Edison felt me all over and said : "Tesla weighs 152 lbs. to an ounce," and he guest it exactly. Stript I weighed 142 lbs. and that is still my weight. I whispered to Mr. Johnson: "How is it possible that Edison could guess my weight so closely?" "Well," he said, lowering his voice, "I will tell you, confidentially, but you must not say anything. He was employed for a long time in a Chicago slaughter-house where he weighed thousands of hogs every day! That's why." My friend, the Hon. Chauncey M. Depew, tells of an Englishman on whom he sprung one of his original anecdotes and who listened with a puzzled expression but—a year later—laughed out loud. I will frankly confess it took me longer than that to appreciate Johnson's joke.

Now, my well being is simply the result of a careful and measured mode of living and perhaps the most astonishing thing is that three times in my youth I was rendered by illness a hopeless physical wreck and given up by physicians. More than this, thru ignorance and lightheartedness, I got into all sorts of difficulties, dangers and scrapes from which I extricated myself as by enchantment. I was almost drowned a dozen times; was nearly boiled alive and just mist being cremated. I was entombed, lost and frozen. I had hair-breadth escapes from mad dogs, hogs, and other wild animals. I past thru dreadful diseases and met with all kinds of odd mishaps and that I am hale and hearty today seems like a miracle. But as I recall these incidents to my mind I feel convinced that my preservation was not altogether accidental.

An inventor's endeavor is essentially life-saving. Whether he harnesses forces, improves devices, or provides new comforts and conveniences, he is adding to the safety of our existence. He is also better qualified than the average individual to protect himself in peril, for he is observant and resourceful. If I had no other evidence that I was, in a measure, possest of such qualities I would find it in these personal experiences. The reader will be able to judge for himself if I mention one or two instances. On one occasion, when about 14 years old, I wanted to scare some friends who were bathing with me. My plan was to dive under a long floating structure and slip out quietly at the other end. Swimming and diving came to me as naturally as to a duck and I was confident that I could perform the feat. Accordingly I plunged into the water and, when out of view, turned around and proceeded rapidly towards the opposite side. Thinking that I was safely beyond the structure, I rose to the surface but to my dismay struck a beam. Of course, I quickly dived and forged ahead with rapid strokes until my breath was beginning to give out. Rising for the second time, my head came again in contact with a beam. Now I was becoming desperate. However, summoning all my energy, I made a third frantic attempt but the result was the same. The torture of supprest breathing was getting unendurable, my brain was reeling and I felt myself sinking. At that moment, when my situation seemed absolutely hopeless, I experienced one of those flashes of light and the structure above me appeared before my vision. I either discerned or guest that there was a little space between the surface of the water and the boards resting on the beams and, with consciousness nearly gone, I floated up, prest my mouth close to the planks and managed to inhale a little air, unfortunately mingled with a spray of water which nearly choked me. Several times I repeated this procedure as in a dream until my heart, which was racing at a terrible rate, quieted down and I gained composure. After that I made a number of unsuccessful dives, having completely lost the sense of direction, but finally succeeded in getting out of the trap when my friends had already given me up and were fishing for my body.

That bathing season was spoiled for me thru recklessness but I soon forgot the lesson and only two years later I fell into a worse predicament. There was a large flour mill with a dam across the river near the city where I was studying at that time. As a rule the height of the water was only two or three inches above the dam and to swim out to it was a sport not very dangerous in which I often indulged. One day I went alone to the river to enjoy myself as usual. When I was a short distance from the masonry, however, I was horrified to observe that the water had risen and was carrying me along swiftly. I tried to get away but it was too late. Luckily, tho, I saved myself from being swept over by taking hold of the wall with both hands. The pressure against my chest was great and I was barely able to keep my head above the surface. Not a soul was in sight and my voice was lost in the roar of the fall. Slowly and gradually I became exhausted and unable to withstand the strain longer. Just as I was about to let go, to be dashed against the rocks below, I saw in a flash of light a familiar diagram illustrating the hydraulic principle that the pressure of a fluid in motion is proportionate to the area exposed, and automatically I turned on my left side. As if by magic the pressure was reduced and I found it comparatively easy in that position to resist the force of the stream. But the danger still confronted me. I knew that sooner or later I would be carried down, as it was not possible for any help to reach me in time, even if I attracted attention. I am ambidextrous now but then I was left-handed and had comparatively little strength in my right arm. For this reason I did not dare to turn on the other side to rest and nothing remained but to slowly push my body along the dam. I had to get away from the mill towards which my face was turned as the current there was much swifter and deeper. It was a long and painful ordeal and I came near to failing at its very end for I was confronted with a depression in the masonry. I managed to get over with the last ounce of my force and fell in a swoon when I reached the bank, where I was found. I had torn virtually all the skin from my left side and it took several weeks before the fever subsided and I was well. These are only two of many instances but they may be sufficient to show that had it not been for the inventor's instinct I would not have lived to tell this tale.

Interested people have often asked me how and when I began to invent. This I can only answer from my present recollection in the light of which the first attempt I recall was rather ambitious for it involved the invention of an *apparatus* and a *method*. In the former I was anticipated but the latter was original. It happened in this way. One of my playmates had come into the possession of a hook and fishing-tackle which created quite an excitement in the village, and the next morning all started out to catch frogs. I was left alone and deserted owing to a quarrel with this boy. I had never seen a real hook and pictured it as something wonderful, endowed with peculiar qualities, and was despairing not to be one of the party. Urged by necessity, I somehow got hold of a piece of soft iron wire, hammered the end to a sharp point between two stones, bent it into shape, and fastened it to a strong string. I then cut a rod, gathered some bait, and went down to the brook where there were frogs in abundance. But I could not catch any and was almost discouraged when it occurred to me to dangle the empty hook in front of a frog sitting on a stump. At first he collapsed but by and by his eyes bulged out and became bloodshot, he swelled to twice his normal size and made a vicious snap at the hook. Immediately I pulled him up. I tried the same thing again and again and the method proved infallible. When my comrades, who in spite of their fine outfit had caught nothing, came to me they were green with envy. For a long time I kept my secret and enjoyed the monopoly but finally yielded to the spirit of Christmas. Every boy could then do the same and the following summer brought disaster to the frogs.

In my next attempt I seem to have acted under the first

instinctive impulse which later dominated me—to harness the energies of nature to the service of man. I did this thru the medium of May-bugs—or June-bugs as they are called in America—which were a veritable pest in that country and sometimes broke the branches of trees by the sheer weight of their bodies. The bushes were black with them. I would attach as many as four of them to a cross-piece, rotably arranged on a thin spindle, and transmit the motion of the same to a large disc and so derive considerable "power." These creatures were remarkably efficient, for once they were started they had no sense to stop and continued whirling for hours and hours and the hotter it was the harder they worked. All went well until a strange boy came to the place. He was the son of a retired officer in the Austrian Army. That urchin ate May-bugs alive and enjoyed them as tho they were the finest blue-point oysters. That disgusting sight terminated my endeavors in this promising field and I have never since been able to touch a May-bug or any other insect for that matter.

After that, I believe, I undertook to take apart and assemble the clocks of my grandfather. In the former operation I was always successful but often failed in the latter. So it came that he brought my work to a sudden halt in a manner not too delicate and it took thirty years before I tackled another clockwork again. Shortly thereafter I went into the manufacture of a kind of pop-gun which comprised a hollow tube, a piston, and two plugs of hemp. When firing the gun, the piston was prest against the stomach and the tube was pushed back quickly with both hands. The air between the plugs was comprest and raised to high temperature and one of them was expelled with a loud report. The art consisted in selecting a tube of the proper taper from the hollow stalks. I did very well with that gun but my activities interfered with the window panes in our house and met with painful discouragement. If I remember rightly, I then took to carving swords from pieces of furniture which I could conveniently obtain. At that time I was under the sway of the Serbian national poetry and full of admiration for the feats of the heroes. I used to spend hours in mowing down my enemies in the form of corn-stalks which ruined the crops and netted me several spankings from my mother. Moreover these were not of the formal kind but the genuine article.

I had all this and more behind me before I was six years old and had past thru one year of elementary school in the village of Smiljan where I was born. At this juncture we moved to the little city of Gospic nearby. This change of residence was like a calamity to me. It almost broke my heart to part from our pigeons, chickens and sheep, and our magnificent flock of geese which used to rise to the clouds in the morning and return from the feeding grounds at sundown in battle formation, so perfect that it would have put a squadron of the best aviators of the present day to shame. In our new house I was but a prisoner, watching the strange people I saw thru the window blinds. My bashfulness was such that I would rather have faced a roaring lion than one of the city dudes who strolled about. But my hardest trial came on Sunday when I had to dress up and attend the service. There I met with an accident, the mere thought of which made my blood curdle like sour milk for years afterwards. It was my second adventure in a church. Not long before I was entombed for a night in an old chapel on an inaccessible mountain which was visited only once a year. It was an awful experience, but this one was worse. There was a wealthy lady in town, a good but pompous woman, who used to come to the church gorgeously painted up and attired with an enormous train and attendants. One Sunday I had just finished ringing the bell in the belfry and rushed downstairs when this grand dame was sweeping out and I jumped on her train. It tore off with a ripping noise which sounded like a salvo of musketry fired by raw recruits. My father was livid with rage. He gave me a gentle slap on the cheek, the only corporal punishment he ever administered to me but I almost feel it now. The embarrassment and confusion that followed are indescribable. I was practically ostracised until something else happened which redeemed me in the estimation of the community.

An enterprising young merchant had organized a fire department. A new fire engine was purchased, uniforms provided and the men drilled for service and parade. The engine was, in reality, a pump to be worked by sixteen men and was beautifully painted red and black. One afternoon the official trial was prepared for and the machine was transported to the river. The entire population turned out to witness the great spectacle. When all the speeches and ceremonies were concluded, the command was given to pump, but not a drop of water came from the nozzle. The professors and experts tried in vain to locate the trouble. The fizzle was complete when I arrived at the scene. My knowledge of the mechanism was nil and I knew next to nothing of air pressure, but instinctively I felt for the suction hose in the water and found that it had collapsed. When I waded in the river and opened it up the water rushed forth and not a few Sunday clothes were spoiled. Archimedes running naked thru the streets of Syracuse and shouting Eureka at the top of his voice did not make a greater impression than myself. I was carried on the shoulders and was the hero of the day.

Upon settling in the city I began a four-years' course in the so-called Normal School preparatory to my studies at the College or *Real-Gymnasium*. During this period my boyish efforts and exploits, as well as troubles, continued. Among other things I attained the unique distinction of champion crow catcher in the country. My method of procedure was extremely simple. I would go in the forest, hide in the bushes, and imitate the call of the bird. Usually I would get several answers and in a short while a crow would flutter down into the shrubbery near me. After that all I needed to do was to throw a piece of cardboard to detract its attention, jump up and grab it before it could extricate itself from the undergrowth. In this way I would capture as many as I desired. But on one occasion something occurred which made me respect them. I had caught a fine pair of birds and was returning home with a friend. When we left the forest, thousands of crows had gathered making a frightful racket. In a few minutes they rose in pursuit and soon enveloped us. The fun lasted until all of a sudden I received a blow on the back of my head which knocked me down. Then they attacked me viciously. I was compelled to release the two birds and was glad to join my friend who had taken refuge in a cave.

In the schoolroom there were a few mechanical models which interested me and turned my attention to water turbines. I constructed many of these and found great pleasure in operating them. How extraordinary was my life an incident may illustrate. My uncle had no use for this kind of pastime and more than once rebuked me. I was fascinated by a description of Niagara Falls I had perused, and pictured in my imagination a big wheel run by the Falls. I told my uncle that I would go to America and carry out this scheme. Thirty years later I saw my ideas carried out at Niagara and marveled at the unfathomable mystery of the mind.

I made all kinds of other contrivances and contraptions but among these the arbalists I produced were the best. My arrows, when shot, disappeared from sight and at close range traversed a plank of pine one inch thick. Thru the continuous tightening of the bows I developed skin on my stomach very much like that of a crocodile and I am often wondering whether it is due to this exercise that I am able even now to digest cobble-stones! Nor can I pass in silence my performances with the sling which would have enabled me to give a stunning exhibit at the Hippodrome. And now I will tell of one of my feats with this antique implement of war which will strain to the utmost the credulity of the reader. I was practicing while walking with my uncle along the river. The sun was setting, the trout were playful and from time to time one would shoot up into the air, its glistening body sharply defined against a projecting rock beyond. Of course any boy might have hit a fish under these propitious conditions but I undertook a much more difficult task and I foretold to my uncle, to the minutest detail, what I intended doing. I was to hurl a stone to meet the fish, press its body against the rock, and cut it in two. It was no sooner said than done. My uncle looked at me almost scared out of his wits and exclaimed *"Vade retro Satanas!"* and it was a few days before he spoke to me again. Other records, however great, will be eclipsed but I feel that I could peacefully rest on my laurels for a thousand years.

PART III

MY LATER ENDEAVORS

*The Discovery of
the Rotating Magnetic Field*

At the age of ten I entered the Real Gymnasium which was a new and fairly well equipt institution. In the department of physics were various models of classical scientific apparatus, electrical and mechanical. The demonstrations and experiments performed from time to time by the instructors fascinated me and were undoubtedly a powerful incentive to invention. I was also passionately fond of mathematical studies and often won the professor's praise for rapid calculation. This was due to my acquired facility of visualizing the figures and performing the operations, not in the usual intuitive manner, but as in actual life. Up to a certain degree of complexity it was absolutely the same to me whether I wrote the symbols on the board or conjured them before my mental vision. But freehand drawing, to which many hours of the course were devoted, was an annoyance I could not endure. This was rather remarkable as most of the members of the family excelled in it. Perhaps my aversion was simply due to the predilection I found in undisturbed thought. Had it not been for a few exceptionally stupid boys, who could not do anything at all, my record would have been the worst. It was a serious handicap as under the then existing educational regime, drawing being obligatory, this deficiency threatened to spoil my whole career and my father had considerable trouble in railroading me from one class to another.

In the second year at that institution I became obsest with the idea of producing continuous motion thru steady air pressure. The pump incident, of which I have told, had set afire my youthful imagination and imprest me with the boundless possibilities of a vacuum. I grew frantic in my desire to harness this inexhaustible energy but for a long time I was groping in the dark. Finally, however, my endeavors crystallized in an invention which was to enable me to achieve what no other mortal ever attempted. Imagine a cylinder freely rotatable on two bearings and partly surrounded by a rectangular trough which fits it perfectly. The open side of the trough is closed by a partition so that the cylindrical segment within the enclosure divides the latter into two compartments entirely separated from each other by air-tight sliding joints. One of these compartments being sealed and once for all exhausted, the other remaining open, a perpetual rotation of the cylinder would result, at least, I thought so. A wooden model was constructed and fitted with infinite care and when I applied the pump on one side and actually observed that there was a tendency to turning, I was delirious with joy. Mechanical flight was the one thing I wanted to accomplish altho still under the discouraging recollection of a bad fall I sustained by jumping with an umbrella from the top of a building. Every day I used to transport myself thru the air to distant regions but could not understand just how I managed to do it. Now I had something concrete—a flying machine with nothing more than a rotating shaft, flapping wings, and—a vacuum of unlimited power! From that time on I made my daily aërial excursions in a vehicle of comfort and luxury as might have befitted King Solomon. It took years before I understood that the atmospheric pressure acted at right angles to the surface of the cylinder and that the slight rotary effort I observed was due to a leak. Tho this knowledge came gradually it gave me a painful shock. I had hardly completed my course at the Real Gymnasium when I was prostrated with a dangerous illness or rather, a score of them, and my condition became so desperate that I was given up by physicians. During this period I was permitted to read constantly, obtaining books from the Public Library which had been neglected and entrusted to me for classification of the works and preparation of the catalogues. One day I was handed a few volumes of new literature unlike anything I had ever read before and so captivating as to make me utterly forget my hopeless state. They were the earlier works of Mark

Twain and to them might have been due the miraculous recovery which followed. Twenty-five years later, when I met Mr. Clements and we formed a friendship between us, I told him of the experience and was amazed to see that great man of laughter burst into tears.

My studies were continued at the higher Real Gymnasium in Carlstadt, Croatia, where one of my aunts resided. She was a distinguished lady, the wife of a Colonel who was an old war-horse having participated in many battles. I never can forget the three years I past at their home. No fortress in time of war was under a more rigid discipline. I was fed like a canary bird. All the meals were of the highest quality and deliciously prepared but short in quantity by a thousand percent. The slices of ham cut by my aunt were like tissue paper. When the Colonel would put something substantial on my plate she would snatch it away and say excitedly to him: "Be careful, Niko is very delicate." I had a voracious appetite and suffered like Tantalus. But I lived in an atmosphere of refinement and artistic taste quite unusual for those times and conditions. The land was low and marshy and malaria fever never left me while there despite of the enormous amounts of quinin I consumed. Occasionally the river would rise and drive an army of rats into the buildings, devouring everything even to the bundles of the fierce paprika. These pests were to me a welcome diversion. I thinned their ranks by all sorts of means, which won me the unenviable distinction of rat-catcher in the community. At last, however, my course was completed, the misery ended, and I obtained the certificate of maturity which brought me to the cross-roads.

During all those years my parents never wavered in their resolve to make me embrace the clergy, the mere thought of which filled me with dread. I had become intensely interested in electricity under the stimulating influence of my Professor of Physics, who was an ingenious man and often demonstrated the principles by apparatus of his own invention. Among these I recall a device in the shape of a freely rotatable bulb, with tin-foil coatings, which was made to spin rapidly when connected to a static machine. It is impossible for me to convey an adequate idea of the intensity of feeling I experienced in witnessing his exhibitions of these mysterious phenomena. Every impression produced a thousand echoes in my mind. I wanted to know more of this wonderful force; I longed for experiment and investigation and resigned myself to the inevitable with aching heart.

Just as I was making ready for the long journey home I received word that my father wished me to go on a shooting expedition. It was a strange request as he had been always strenuously opposed to this kind of sport. But a few days later I learned that the cholera was raging in that district and, taking advantage of an opportunity, I returned to Gospic in disregard of my parents' wishes. It is incredible how absolutely ignorant people were as to the causes of this scourge which visited the country in intervals of from fifteen to twenty years. They thought that the deadly agents were transmitted thru the air and filled it with pungent odors and smoke. In the meantime they drank the infected water and died in heaps. I contracted the awful disease on the very day of my arrival and altho surviving the crisis, I was confined to bed for nine months with scarcely any ability to move. My energy was completely exhausted and for the second time I found myself at death's door. In one of the sinking spells which was thought to be the last, my father rushed into the room. I still see his pallid face as he tried to cheer me in tones belying his assurance. "Perhaps," I said, "I may get well if you will let me study engineering." "You will go to the best technical institution in the world." he solemnly replied, and I knew that he meant it. A heavy weight was lifted from my mind but the relief would have come too late had it not been for a marvelous cure brought about thru a bitter decoction of a peculiar bean. I came to life like another Lazarus to the utter amazement of everybody. My father insisted that I spend a year in healthful physical outdoor exercises to which I reluctantly consented. For most of this term I roamed in the mountains, loaded with a hunter's outfit and a bundle of books, and this contact with nature made me stronger in body as well as in mind. I thought and planned, and conceived many ideas almost as a rule delusive. The vision was clear enough but the knowledge of principles was very limited. In one of my inventions I proposed to convey letters and packages across the seas, thru a submarine tube, in spherical containers of sufficient strength to resist the hydraulic pressure. The pumping plant, intended to force the water thru the tube, was accurately figured and designed and all other particulars carefully worked out. Only one trifling detail, of no consequence, was lightly dismist. I assumed an arbitrary velocity of the water and, what is more, took pleasure in making it high, thus arriving at a stupendous performance supported by faultless calculations. Subsequent reflections, however, on the resistance of pipes to fluid flow determined me to make this invention public property.

Another one of my projects was to construct a ring around the equator which would, of course, float freely and could be arrested in its spinning motion by reactionary forces, thus enabling travel at a rate of about one thousand miles an hour, impracticable by rail. The reader will smile. The plan was difficult of execution, I will admit, but not nearly so bad as that of a well-known New York professor, who wanted to pump the air from the torrid to the temperate zones, entirely forgetful of the fact that the Lord had provided a gigantic machine for this very purpose.

Still another scheme, far more important and attractive, was to derive power from the rotational energy of terrestrial bodies. I had discovered that objects on the earth's surface, owing to the diurnal rotation of the globe, are carried by the same alternately in and against the direction of translatory movement. From this results a great change in momentum which could be utilized in the simplest imaginable manner to furnish motive effort in any habitable region of the world. I cannot find words to describe my disappointment when later I realized that I was in the predicament of Archimedes, who vainly sought for a fixt point in the universe.

At the termination of my vacation I was sent to the Polytechnic School in Gratz, Styria, which my father had chosen as one of the oldest and best reputed institutions. That was the moment I had eagerly awaited and I began my studies under

good auspices and firmly resolved to succeed. My previous training was above the average, due to my father's teaching and
opportunities afforded. I had acquired the knowledge of a
number of languages and waded thru the books of several
libraries, picking up information more or less useful. Then
again, for the first time, I could choose my subjects as I liked,
and free-hand drawing was to bother me no more. I had made
up my mind to give my parents a surprise, and during the
whole first year I regularly started my work at three o'clock in
the morning and continued until eleven at night, no Sundays
or holidays excepted. As most of my fellow-students took things
easily, naturally enough I eclipsed all records. In the course of
that year I past thru nine exams and the professors thought I
deserved more than the highest qualifications. Armed with their
flattering certificates, I went home for a short rest, expecting a
triumph, and was mortified when my father made light of these
hard-won honors. That almost killed my ambition; but later,
after he had died, I was pained to find a package of letters
which the professors had written him to the effect that unless
he took me away from the Institution I would be killed thru
overwork. Thereafter I devoted myself chiefly to physics, mechanics and mathematical studies, spending the hours of
leisure in the libraries. I had a veritable mania for finishing
whatever I began, which often got me into difficulties. On one
occasion I started to read the works of Voltaire when I learned,
to my dismay, that there were close on one hundred large volumes in small print which that monster had written while
drinking seventy-two cups of black coffee per diem. It had to
be done, but when I laid aside the last book I was very glad,
and said, "Never more!"

My first year's showing had won me the appreciation and
friendship of several professors. Among these were Prof. Rogner,
who was teaching arithmetical subjects and geometry; Prof.
Poeschl, who held the chair of theoretical and experimental
physics, and Dr. Allé, who taught integral calculus and spe-
cialized in differential equations. This scientist was the most
brilliant lecturer to whom I ever listened. He took a special
interest in my progress and would frequently remain for an

hour or two in the lecture room, giving me problems to solve, in which I delighted. To him I explained a flying machine I had conceived, not an illusionary invention, but one based on sound, scientific principles, which has become realizable thru my turbine and will soon be given to the world. Both Professors Rogner and Poeschl were curious men. The former had peculiar ways of expressing himself and whenever he did so there was a riot, followed by a long and embarrassing pause. Prof. Poeschl was a methodical and thoroly grounded German. He had enormous feet and hands like the paws of a bear, but all of his experiments were skillfully performed with clock-like precision and without a miss.

It was in the second year of my studies that we received a Gramme dynamo from Paris, having the horseshoe form of a laminated field magnet, and a wire-wound armature with a commutator. It was connected up and various effects of the currents were shown. While Prof. Poeschl was making demonstrations, running the machine as a motor, the brushes gave trouble, sparking badly, and I observed that it might be possible to operate a motor without these appliances. But he declared that it could not be done and did me the honor of delivering a lecture on the subject, at the conclusion of which he remarked: "Mr. Tesla may accomplish great things, but he certainly never will do this. It would be equivalent to converting a steadily pulling force, like that of gravity, into a rotary effort. It is a perpetual motion scheme, an impossible idea." But instinct is something which transcends knowledge. We have, undoubtedly, certain finer fibers that enable us to perceive truths when logical deduction, or any other willful effort of the brain, is futile. For a time I wavered, imprest by the professor's authority, but soon became convinced I was right and undertook the task with all the fire and boundless confidence of youth.

I started by first picturing in my mind a direct-current machine, running it and following the changing flow of the currents in the armature. Then I would imagine an alternator and investigate the processes taking place in a similar manner. Next I would visualize systems comprising motors and generators and operate them in various ways. The images I saw were to me perfectly real and tangible. All my remaining term in Gratz was past in intense but fruitless efforts of this kind, and I almost came to the conclusion that the problem was insolvable. In 1880 I went to Prague, Bohemia, carrying out my father's wish to complete my education at the University there. It was in that city that I made a decided advance, which consisted in detaching the commutator from the machine and studying the phenomena in this new aspect, but still without result. In the year following there was a sudden change in my views of life. I realized that my parents had been making too great sacrifices on my account and resolved to relieve them of the burden. The wave of the American telephone had just reached the European continent and the system was to be installed in Budapest, Hungary. It appeared an ideal opportunity, all the more as a friend of our family was at the head of the enterprise. It was here that I suffered the complete breakdown of the nerves to which I have referred. What I experienced during the period of that illness surpasses all belief. My sight and hearing were always extraordinary. I could clearly discern objects in the distance when others saw no trace of them. Several times in my boyhood I saved the houses of our neighbors from fire by hearing the faint crackling sounds which did not disturb their sleep, and calling for help.

In 1899, when I was past forty and carrying on my experiments in Colorado, I could hear very distinctly thunderclaps at a distance of 550 miles. The limit of audition for my young assistants was scarcely more than 150 miles. My ear was thus over thirteen times more sensitive. Yet at that time I was, so to speak, stone deaf in comparison with the acuteness of my hearing while under the nervous strain. In Budapest I could hear the ticking of a watch with three rooms between me and the time-piece. A fly alighting on a table in the room would cause a dull thud in my ear. A carriage passing at a distance of a few miles fairly shook my whole body. The whistle of a locomotive twenty or thirty miles away made the bench or chair on which I sat vibrate so strongly that the pain was unbearable. The ground under my feet trembled continuously. I had to support

my bed on rubber cushions to get any rest at all. The roaring
noises from near and far often produced the effect of spoken
words which would have frightened me had I not been able to
resolve them into their accidental components. The sun's rays,
when periodically intercepted, would cause blows of such force
on my brain that they would stun me. I had to summon all my
will power to pass under a bridge or other structure as I experienced a crushing pressure on the skull. In the dark I had the
sense of a bat and could detect the presence of an object at a
distance of twelve feet by a peculiar creepy sensation on the
forehead. My pulse varied from a few to two hundred and
sixty beats and all the tissues of the body with twitchings and
tremors which was perhaps the hardest to bear. A renowned
physician who gave me daily large doses of Bromid of Potas-
sium pronounced my malady unique and incurable. It is my
eternal regret that I was not under the observation of experts
in physiology and psychology at that time. I clung desperately
to life, but never expected to recover. Can anyone believe that
so hopeless a physical wreck could ever be transformed into a
man of astonishing strength and tenacity, able to work thirty-
eight years almost without a day's interruption, and find himself still strong and fresh in body and mind? Such is my case. A
powerful desire to live and to continue the work, and the
assistance of a devoted friend and athlete accomplished the
wonder. My health returned and with it the vigor of mind. In
attacking the problem again I almost regretted that the struggle was soon to end. I had so much energy to spare. When I
undertook the task it was not with a resolve such as men often
make. With me it was a sacred vow, a question of life and death.
I knew that I would perish if I failed. Now I felt that the battle
was won. Back in the deep recesses of the brain was the solution, but I could not yet give it outward expression. One afternoon, which is ever present in my recollection, I was enjoying
a walk with my friend in the City Park and reciting poetry.
At that age I knew entire books by heart, word for word. One
of these was Göethe's "Faust." The sun was just setting and
reminded me of the glorious passage:

"Sie rückt und weicht, der Tag ist überlebt, Dort eilt sie hin und fördert neues Leben. Oh, dass kein Flügel mich vom Boden hebt Ihr nach und immer nach zu streben!*

Ein schöner Traum indessen sie entweicht, Ach, zu des Geistes Flügeln wird so leicht Kein körperlicher Flügel sich gesellen!"†

As I uttered these inspiring words the idea came like a flash of lightning and in an instant the truth was revealed. I drew with a stick on the sand the diagrams shown six years later in my address before the American Institute of Electrical Engineers, and my companion understood them perfectly. The images I saw were wonderfully sharp and clear and had the solidity of metal and stone, so much so that I told him: "See my motor here; watch me reverse it." I cannot begin to describe my emotions. Pygmalion seeing his statue come to life could not have been more deeply moved. A thousand secrets of nature which I might have stumbled upon accidentally I would have given for that one which I had wrested from her against all odds and at the peril of my existence.

---

* "The glow retreats, done is the day of toil;
       It yonder hastes, new fields of life exploring;
       Ah, that no wing can lift me from the soil,
       Upon its track to follow, follow soaring!"

†   A glorious dream! though now the glories fade.
       Alas! the wings that lift the mind no aid
       Of wings to lift the body can bequeath me."

PART IV

THE DISCOVERY OF
THE TESLA COIL AND
TRANSFORMER

For a while I gave myself up entirely to the intense enjoyment of picturing machines and devising new forms. It was a mental state of happiness about as complete as I have ever known in life. Ideas came in an uninterrupted stream and the only difficulty I had was to hold them fast. The pieces of apparatus I conceived were to me absolutely real and tangible in every detail, even to the minutest marks and signs of wear. I delighted in imagining the motors constantly running, for in this way they presented to the mind's eye a more fascinating sight. When natural inclination develops into a passionate desire, one advances towards his goal in seven-league boots. In less than two months I evolved virtually all the types of motors and modifications of the system which are now identified with my name. It was, perhaps, providential that the necessities of existence commanded a temporary halt to this consuming activity of the mind. I came to Budapest prompted by a premature report concerning the telephone enterprise and, as irony of fate willed it, I had to accept a position as draftsman in the Central Telegraph Office of the Hungarian Government at a salary which I deem it my privilege not to disclose! Fortunately, I soon won the interest of the Inspector-in-Chief and was thereafter employed on calculations, designs and estimates in connection with new installations, until the Telephone Exchange was started, when I took charge of the same. The knowledge and practical experience I gained in the course of this work was most valuable and the employment gave me ample opportunities for the exercise of my inventive faculties. I made several improvements in the Central Station apparatus and perfected a telephone repeater or amplifier which was never patented or publicly described but would be creditable to me even today. In recognition of my efficient assistance the organizer of the undertaking, Mr. Puskas, upon disposing of his business in Budapest, offered me a position in Paris which I gladly accepted.

I never can forget the deep impression that magic city produced on my mind. For several days after my arrival I roamed thru the streets in utter bewilderment of the new spectacle. The attractions were many and irresistible, but, alas, the income was spent as soon as received. When Mr. Puskas asked me how I was getting along in the new sphere, I described the situation accurately in the statement that "the last twenty-nine days of the month are the toughest!" I led a rather strenuous life in what would now be termed "Rooseveltian fashion." Every morning, regardless of weather, I would go from the Boulevard St. Marcel, where I resided, to a bathing house on the Seine, plunge into the water, loop the circuit twenty-seven times and then walk an hour to reach Ivry, where the Company's factory was located. There I would have a wood-chopper's breakfast at half-past seven o'clock and then eagerly await the lunch hour, in the meanwhile cracking hard nuts for the Manager of the Works, Mr. Charles Batchellor, who was an intimate friend and assistant of Edison. Here I was thrown in contact with a few Americans who fairly fell in love with me because of my proficiency in—billiards. To these men I explained my invention and one of them, Mr. D. Cunningham, Foreman of the Mechanical Department, offered to form a stock company. The proposal seemed to me comical in the extreme. I did not have the faintest conception of what that meant except that it was an American way of doing things. Nothing came of it, however, and during the next few months I had to travel from one to another place in France and Germany to cure the ills of the power plants. On my return to Paris I submitted to one of the administrators of the Company, Mr. Rau, a plan for improving their dynamos and was given an opportunity. My success was complete and the delighted directors accorded me the privilege of developing automatic regulators which were much desired. Shortly after there was some trouble with the lighting plant which has been installed at the new railroad station in Strassburg, Alsace. The wiring was defective and on the occasion of the opening ceremonies a large part of a wall was blown out thru a short-circuit right in the presence of old Emperor William I. The German Government refused to take the plant and the French Company was facing a serious loss. On account of my knowledge of the German language and past experience, I was entrusted with the difficult task of straightening out matters and early in 1883 I went to Strassburg on that mission.

The First Induction Motor Is Built

Some of the incidents in that city have left an indelible record on my memory. By a curious coincidence, a number of men who subsequently achieved fame, lived there about that time. In later life I used to say, "There were bacteria of greatness in that old town. Others caught the disease but I escaped!" The practical work, correspondence, and conferences with officials kept me preoccupied day and night, but as soon as I was able to manage I undertook the construction of a simple motor in a mechanical shop opposite the railroad station, having brought with me from Paris some material for that purpose. The consummation of the experiment was, however, delayed until the summer of that year when I finally had the satisfaction of *seeing rotation effected by alternating currents of different phase, and without sliding contacts or commutator,* as I had conceived a year before. It was an exquisite pleasure but not to compare with the delirium of joy following the first revelation.

Among my new friends was the former Mayor of the city, Mr. Bauzin, whom I had already in a measure acquainted with this and other inventions of mine and whose support I endeavored to enlist. He was sincerely devoted to me and put my project before several wealthy persons but, to my mortification, found no response. He wanted to help me in every possible way and the approach of the first of July, 1919, happens to remind me of a form of "assistance" I received from that charming man, which was not financial but none the less appreciated. In 1870, when the Germans invaded the country, Mr. Bauzin had buried a good sized allotment of St. Estèphe of 1801 and he came to the conclusion that he knew no worthier person than myself to consume that precious beverage. This, I may say, is one of the unforgettable incidents to which I have referred. My friend urged me to return to Paris as soon as possible and seek support there. This I was anxious to do but my work and negotiations were protracted owing to all sorts of petty obstacles I encountered so that at times the situation seemed hopeless.

German "Efficiency"

Just to give an idea of German thoroness and "efficiency," I may mention here a rather funny experience. An incandescent lamp of 16 c.p. was to be placed in a hallway and upon selecting the proper location I ordered the *monteur* to run the wires. After working for a while he concluded that the engineer had to be consulted and this was done. The latter made several objections but ultimately agreed that the lamp should be placed two inches from the spot I had assigned, whereupon the work proceeded. Then the engineer became worried and told me that Inspector Averdeck should be notified. That important person called, investigated, debated, and decided that the lamp should be shifted back two inches, which was the place I had marked. It was not long, however, before Averdeck got cold feet himself and advised me that he had informed *Ober-Inspector* Hieronimus of the matter and that I should await his decision. It was several days before the *Ober-Inspector* was able to free himself of other pressing duties but at last he arrived and a two-hour debate followed, when he decided to move the lamp two inches farther. My hopes that this was the final act were shattered when the *Ober-Inspector* returned and said to me: "*Regierungsrath* Funke is so particular that I would not dare to give an order for placing this lamp without his explicit approval." Accordingly arrangements for a visit from that great man were made. We started cleaning up and polishing early in the morning. Everybody brushed up, I put on my gloves and when Funke came with his retinue he was ceremoniously received. After two hours' deliberation he suddenly exclaimed: "I must be going," and pointing to a place on the ceiling, he ordered me to put the lamp there. It was the exact spot which I had originally chosen.

So it went day after day with variations, but I was determined to achieve at whatever cost and in the end my efforts were rewarded. By the spring of 1884 all the differences were adjusted, the plant formally accepted, and I returned to Paris with pleasing anticipations. One of the administrators had promised me a liberal compensation in case I succeeded, as well as a fair consideration of the improvements I had made in their dynamos and I hoped to realize a substantial sum. There were three administrators whom I shall designate as A, B and C for convenience. When I called on A *he* told me that B had the say. This gentleman thought that only C could decide and the latter was quite sure that A alone had the power to act. After several laps of this *circulus viciosus*, it dawned upon me that my reward was a castle in Spain. The utter failure of my attempts to raise capital for development was another disappointment and when Mr. Batchellor prest me to go to America with a view of redesigning the Edison machines, I determined to try my fortunes in the Land of Golden Promise. But the chance was nearly mist. I liquefied my modest assets, secured accommodations and found myself at the railroad station as the train was pulling out. At that moment I discovered that my money and tickets were gone. What to do was the question. Hercules had plenty of time to deliberate but I had to decide while running alongside the train with opposite feelings surging in my brain like condenser oscillations. Resolve, helped by dexterity, won out in the nick of time and upon passing thru the usual experiences, as trivial as unpleasant, I managed to embark for New York with the remnants of my belongings, some poems and articles I had written, and a package of calculations relating to solutions of an unsolvable integral and to my flying machine. During the voyage I sat most of the time at the stern of the ship watching for an opportunity to save somebody from a watery grave, without the slightest thought of danger. Later when I had absorbed some of the practical American sense I shivered at the recollection and marvelled at my former folly.

Tesla in America

I wish that I could put in words my first impressions of this country. In the Arabian Tales I read how genii transported people into a land of dreams to live thru delightful adventures. My case was just the reverse. The genii had carried me from a world of dreams into one of realities. What I had left was beautiful, artistic and fascinating in every way; what I saw here was machined, rough and unattractive. A burly policeman was twirling his stick which looked to me as big as a log. I approached him politely with the request to direct me. "Six blocks down, then to the left," he said, with murder in his eyes. "Is this America?" I asked myself in painful surprise. "It is a century behind Europe in civilization." When I went abroad in 1889— five years having elapsed since my arrival here—I became convinced that *it was more than one hundred years AHEAD of Europe* and nothing has happened to this day to change my opinion.

Tesla Meets Edison

The meeting with Edison was a memorable event in my life. I was amazed at this wonderful man who, without early advantages and scientific training, had accomplished so much. I had studied a dozen languages, delved in literature and art, and had spent my best years in libraries reading all sorts of stuff that fell into my hands, from Newton's *"Principia"* to the novels of Paul de Kock, and felt that most of my life had been squandered. But it did not take long before I recognized that it was the best thing I could have done. Within a few weeks I had won Edison's confidence and it came about in this way.

The S. S. *Oregon*, the fastest passenger steamer at that time, had both of its lighting machines disabled and its sailing was delayed. As the superstructure had been built after their installation it was impossible to remove them from the hold. The predicament was a serious one and Edison was much annoyed. In the evening I took the necessary instruments with me and went aboard the vessel where I stayed for the night. The dynamos were in bad condition, having several short-circuits and breaks, but with the assistance of the crew I succeeded in putting them in good shape. At five o'clock in the morning, when passing along Fifth Avenue on my way to the shop, I met Edison with Batchellor and a few others as they were returning home to retire. "Here is our Parisian running around at night," he said. When I told him that I was coming from the *Oregon* and had repaired both machines, he looked at me in silence and walked away without another word. But when he had gone some distance I heard him remark: "Batchellor, this is a d—n good man," and from that time on I had full freedom in directing the work. For nearly a year my regular hours were from 10.30 A. M. until 5 o'clock the next morning without a day's exception. Edison said to me: "I have had many hard-working assistants but you take the cake." During this period I designed twenty-four different types of standard machines with short cores and of uniform pattern which replaced the old ones. The Manager had promised me fifty thousand dollars on the completion of this task but it turned out to be a practical joke. This gave me a painful shock and I resigned my position.

Immediately thereafter some people approached me with the proposal of forming an arc light company under my name, to which I agreed. Here finally was an opportunity to develop the motor, but when I broached the subject to my new associates they said: "No, we want the arc lamp. We don't care for this alternating current of yours." In 1886 my system of arc lighting was perfected and adopted for factory and municipal lighting, and I was free, but with no other possession than a beautifully engraved certificate of stock of hypothetical value. Then followed a period of struggle in the new medium for which I was not fitted, but the reward came in the end and in April, 1887, the Tesla Electric Company was organized, providing a laboratory and facilities. The motors I built there were exactly as I had imagined them. I made no attempt to improve the design, but merely reproduced the pictures as they appeared to my vision and the operation was always as I expected.

In the early part of 1888 an arrangement was made with the Westinghouse Company for the manufacture of the motors on a large scale. But great difficulties had still to be overcome. My system was based on the use of low frequency currents and the Westinghouse experts had adopted 133 cycles with the object of securing advantages in the transformation. They did not want to depart from their standard forms of apparatus and my efforts had to be concentrated upon adapting the motor to these conditions. Another necessity was to produce a motor capable of running efficiently at this frequency on two wires which was not easy of accomplishment.

At the close of 1889, however, my services in Pittsburg being no longer essential, I returned to New York and resumed experimental work in a laboratory on Grand Street, where I began immediately the design of high frequency machines. The problems of construction in this unexplored field were novel and quite peculiar and I encountered many difficulties. I rejected the inductor type, fearing that it might not yield perfect sine waves which were so important to resonant action. Had it not been for this I could have saved myself a great deal of labor. Another discouraging feature of the high frequency alternator seemed to be the inconstancy of speed which threatened to impose serious limitations to its use. I had already noted in my demonstrations before the American Institution of Electrical Engineers that several times the tune was lost, necessitating readjustment, and did not yet foresee, what I discovered long afterwards, a means of operating a machine of this kind at a speed constant to such a degree as not to vary more than a small fraction of one revolution between the extremes of load.

The Invention of the Tesla Coil

From many other considerations it appeared desirable to invent a simpler device for the production of electric oscillations. In 1856 Lord Kelvin had exposed the theory of the condenser discharge, but no practical application of that important knowledge was made. I saw the possibilities and undertook the development of induction apparatus on this principle. My progress was so rapid as to enable me to exhibit at my lecture in 1891 a coil giving sparks of *five inches*. On that occasion I frankly told the engineers of a defect involved in the transformation by the new method, namely, the loss in the spark gap. Subsequent investigation showed that no matter what medium is employed, be it air, hydrogen, mercury vapor, oil or a stream of electrons, the efficiency is the same. It is a law very much like that governing the conversion of mechanical energy. We may drop a weight from a certain height vertically down or carry it to the lower level along any devious path, it is immaterial insofar as the amount of work is concerned. Fortunately however, this drawback is not fatal as by proper proportioning of the resonant circuits an *efficiency of 85 per cent* is attainable. Since my early announcement of the invention it has come into universal use and wrought a revolution in many departments. But a still greater future awaits it. When in 1900 I obtained powerful discharges of 100 feet and flashed a current around the globe, I was reminded of the first tiny spark I observed in my Grand Street laboratory and was thrilled by sensations akin to those I felt when I discovered the *rotating magnetic field*.

PART V

THE MAGNIFYING
TRANSMITTER

As I review the events of my past life I realize how subtle are the influences that shape our destinies. An incident of my youth may serve to illustrate. One winter's day I managed to climb a steep mountain, in company with other boys. The snow was quite deep and a warm southerly wind made it just suitable for our purpose. We amused ourselves by throwing balls which would roll down a certain distance, gathering more or less snow, and we tried to outdo one another in this exciting sport. Suddenly a ball was seen to go beyond the limit, swelling to enormous proportions until it became as big as a house and plunged thundering into the valley below with a force that made the ground tremble. I looked on spellbound, incapable of understanding what had happened. For weeks afterward the picture of the avalanche was before my eyes and I wondered how anything so small could grow to such an immense size. Ever since that time the magnification of feeble actions fascinated me, and when, years later, I took up the experimental study of mechanical and electrical resonance, I was keenly interested from the very start. Possibly, had it not been for that early powerful impression, I might not have followed up the little spark I obtained with my coil and never developed my best invention, the true history of which I will tell here for the first time.

Scrapping the World's Engines

"Lionhunters" have often asked me which of my discoveries I prize most. This depends on the point of view. Not a few tech-

nical men, very able in their special departments, but dominated by a pedantic spirit and nearsighted, have asserted that
excepting the induction motor I have given to the world little
of practical use. This is a grievous mistake. A new idea must
not be judged by its immediate results. My alternating system
of power transmission came at a psychological moment, as a
long-sought answer to pressing industrial questions, and altho
considerable resistance had to be overcome and opposing
interests reconciled, as usual, the commercial introduction
could not be long delayed. Now, compare this situation with
that confronting my turbine, for example. One should think
that so simple and beautiful an invention, possessing many
features of an ideal motor, should be adopted at once and,
undoubtedly, it would under similar conditions. But the pro-
spective effect of the rotating field was not to render worthless
existing machinery; on the contrary, it was to give it additional
value. The system lent itself to new enterprise as well as to
improvement of the old. My turbine is an advance of a character entirely different. It is a radical departure in the sense that
its success would mean the abandonment of the antiquated
types of prime movers on which billions of dollars have been
spent. Under such circumstances the progress must needs be
slow and perhaps the greatest impediment is encountered in
the prejudicial opinions created in the minds of experts by
organized opposition. Only the other day I had a disheartening
experience when I met my friend and former assistant, Charles
F. Scott, now professor of Electrical Engineering at Yale. I had
not seen him for a long time and was glad to have an opportunity for a little chat at my office. Our conversation naturally
enough drifted on my turbine and I became heated to a high
degree. "Scott," I exclaimed, carried away by the vision of a
glorious future, "my turbine will scrap all the heat-engines in
the world." Scott stroked his chin and looked away thought-
fully, as though making a mental calculation. "That will make
quite a pile of scrap," he said, and left without another word!

"Aladdin's Lamp"

These and other inventions of mine, however, were nothing more than steps forward in certain directions. In evolving them I simply followed the inborn instinct to improve the present devices without any special thought of our far more imperative necessities. The "Magnifying Transmitter" was the product of labors extending through years, having for their chief object the solution of problems which are infinitely more important to mankind than mere industrial development.

If my memory serves me right, it was in November, 1890, that I performed a laboratory experiment which was one of the most extraordinary and spectacular ever recorded in the annals of science. In investigating the behaviour of high frequency currents I had satisfied myself that an electric field of sufficient intensity could be produced in a room to light up electrodeless vacuum tubes. Accordingly, a transformer was built to test the theory and the first trial proved a marvelous success. It is difficult to appreciate what those strange phenomena meant at that time. We crave for new sensations but soon become indifferent to them. The wonders of yesterday are today common occurrences. When my tubes were first publicly exhibited they were viewed with amazement impossible to describe. From all parts of the world I received urgent invitations and numerous honors and other flattering inducements were offered to me, which I declined.

In Faraday's Chair

But in 1892 the demands became irresistible and I went to London where I delivered a lecture before the Institution of Electrical Engineers. It had been my intention to leave immediately for Paris in compliance with a similar obligation, but Sir James Dewar insisted on my appearing before the Royal Institution. I was a man of firm resolve but succumbed easily to the forceful arguments of the great Scotchman. He pushed me into a chair and poured out half a glass of a wonderful brown fluid which sparkled in all sorts of iridescent colors and tasted like nectar. "Now," said he, "you are sitting in Faraday's chair and you are enjoying whiskey he used to drink." In both aspects it was an enviable experience. The next evening I gave a demonstration before that Institution, at the termination of which Lord Rayleigh addressed the audience and his generous words gave me the first start in these endeavors. I fled from London and later from Paris to escape favors showered upon me, and journeyed to my home where I passed through a most painful ordeal and illness. Upon regaining my health I began to formulate plans for the resumption of work in America. Up to that time I never realized that I possessed any particular gift of discovery but Lord Rayleigh, whom I always considered as an ideal man of science, had said so and if that was the case I felt that I should concentrate on some big idea.

Nature's Trigger

One day, as I was roaming in the mountains, I sought shelter from an approaching storm. The sky became overhung with heavy clouds but somehow the rain was delayed until, all of a sudden, there was a lightning flash and a few moments after a deluge. This observation set me thinking. It was manifest that the two phenomena were closely related, as cause and effect, and a little reflection led me to the conclusion that the electrical energy involved in the precipitation of the water was inconsiderable, the function of lightning being much like that of a sensitive trigger. Here was a stupendous possibility of achievement. If we could produce electric effects of the required quality, this whole planet and the conditions of existence on it could be transformed. The sun raises the water of the oceans and winds drive it to distant regions where it remains in a state of most delicate balance. If it were in our power to upset it when and wherever desired, this mighty life-sustaining stream could be at will controlled. We could irrigate arid deserts, create lakes and rivers and provide motive power in unlimited amounts. This would be the most efficient way of harnessing the sun to the uses of man. The consummation depended on our ability to develop electric forces of the order of those in nature. It seemed a hopeless undertaking, but I made up my mind to try it and immediately on my return to the United States, in the summer of 1892, work was begun which was to me all the more attractive, because a means of the same kind was necessary for the successful transmission of energy without wires.

Four Million Volts

The first gratifying result was obtained in the spring of the succeeding year when I reached tensions of about 1,000,000 volts with my conical coil. That was not much in the light of the present art, but it was then considered a feat. Steady progress was made until the destruction of my laboratory by fire in 1895, as may be judged from an article by T. C. Martin which appeared in the April number of the *Century Magazine*. This calamity set me back in many ways and most of that year had to be devoted to planning and reconstruction. However, as soon as circumstances permitted, I returned to the task. Although I knew that higher electro-motive forces were attainable with apparatus of larger dimensions, I had an instinctive perception that the object could be accomplished by the proper design of a comparatively small and compact transformer. In carrying on tests with a *secondary in the form of a flat spiral*, as illustrated in my patents, the absence of streamers surprised me, and it was not long before I discovered that this was due to the position of the turns and their mutual action. Profiting from this observation I resorted to the use of a high tension conductor with turns of considerable diameter sufficiently separated to keep down the distributed capacity, while at the same time preventing undue accumulation of the charge at any point. The application of this principle enabled me to produce pressures of 4,000,000 volts, which was about the limit obtainable in my new laboratory at Houston Street, as the discharges extended through a distance of 16 feet. A photograph of this transmitter was published in the *Electrical Review* of

November, 1898. In order to advance further along this line I had to go into the open, and in the spring of 1899, having completed preparations for the erection of a wireless plant, I went to Colorado where I remained for more than one year. Here I introduced other improvements and refinements which made it possible to generate currents of any tension that may be desired. Those who are interested will find some information in regard to the experiments I conducted there in my article, "The Problem of Increasing Human Energy" in the *Century Magazine* of June, 1900, to which I have referred on a previous occasion.

The Magnifying Transmitter

I have been asked by the ELECTRICAL EXPERIMENTER to be quite explicit on this subject so that my young friends among the readers of the magazine will clearly understand the construction and operation of my "Magnifying Transmitter" and the purposes for which it is intended. Well, then, in the first place, it is a *resonant transformer* with a secondary in which the parts, charged to a high potential, are of considerable area and arranged in space along ideal enveloping surfaces of very large radii of curvature, and at proper distances from one another thereby insuring a *small electric surface density everywhere* so that *no leak can occur even if the conductor is bare.* It is suitable for any frequency, from a few to many thousands of cycles per second, and can be used in the production of currents of tremendous volume and moderate pressure, or of smaller amperage and immense electro-motive force. The maximum electric *tension is merely dependent on the curvature of the surfaces* on which the charged elements are situated and the area of the latter.

100 *Million Volts Possible*

Judging from my past experience, as much as 100,000,000 volts are perfectly practicable. On the other hand currents of many thousands of amperes may be obtained in the antenna. A plant of but very moderate dimensions is required for such performances. Theoretically, a terminal of less than 90 feet in diameter is sufficient to develop an electro-motive force of that magnitude while for antenna currents of from 2,000–4,000 amperes at the usual frequencies it need not be larger than 30 feet in diameter.

In a more restricted meaning this wireless transmitter is one in which the Hertz-wave radiation is an entirely negligible quantity as compared with the whole energy, under which condition the damping factor is extremely small and an enormous charge is stored in the elevated capacity. Such a circuit may then be excited with impulses of any kind, even of low frequency and it will yield sinusoidal and continuous oscillations like those of an alternator.

Taken in the narrowest significance of the term, however, it is a resonant transformer which, besides possessing these qualities, is accurately proportioned to fit the globe and its electrical constants and properties, by virtue of which design it becomes highly efficient and effective in the wireless transmission of energy. Distance is then absolutely eliminated, there being *no diminution in the intensity of the transmitted impulses.* It is even possible to make the actions *increase with the distance from the plant* according to an exact mathematical law.

This invention was one of a number comprised in my "World-System" of wireless transmission which I undertook to commercialize on my return to New York in 1900. As to the immediate purposes of my enterprise, they were clearly outlined in a technical statement of that period from which I quote:

"The 'World-System' has resulted from a combination of several original discoveries made by the inventor in the course of long continued research and experimentation. It makes possible not only the instantaneous and precise wireless transmission of any kind of signals, messages or characters, to all parts of the world, but also the inter-connection of the existing telegraph, telephone, and other signal stations without any change in their present equipment. By its means, for instance, a telephone subscriber here may call up and talk to any other subscriber on the Globe. An inexpensive receiver, not bigger than a watch, will enable him to listen anywhere, on land or sea, to a speech delivered or music played in some other place, however distant. These examples are cited merely to give an idea of the possibilities of this great scientific advance, which annihilates distance and makes that perfect natural conductor, the Earth, available for all the innumerable purposes which human ingenuity has found for a line-wire. One far-reaching result of this is that any device capable of being operated thru one or more wires (at a distance obviously restricted) can likewise be actuated, without artificial conductors and with the same facility and accuracy, at distances to which there are no limits other than those imposed by the physical dimensions of the Globe. Thus, not only will entirely new fields for commercial exploitation be opened up by this ideal method of transmission but the old ones vastly extended.

"The 'World-System' is based on the application of the following important inventions and discoveries:

"1. *The 'Tesla Transformer.'* This apparatus is in the production of electrical vibrations as revolutionary as gunpowder was in warfare. Currents many times stronger than any ever generated in the usual ways, and sparks over one hundred feet long, have been produced by the inventor with an instrument of this kind.

"2. *The 'Magnifying Transmitter.'* This is Tesla's best invention—a peculiar transformer specially adapted to excite the Earth, which is in the transmission of electrical energy what the telescope is in astronomical observation. By the use of this marvelous device he has already set up electrical movements of greater intensity than those of lightning and passed a current, sufficient to light more than two hundred incandescent lamps, around the Globe.

"3. *The 'Tesla Wireless System.'* This system comprises a number of improvements and is the only means known for transmitting economically electrical energy to a distance without wires. Careful tests and measurements in connection with an experimental station of great activity, erected by the inventor in Colorado, have demonstrated that power in any desired amount can be conveyed, clear across the Globe if necessary, with a loss not exceeding a few per cent.

"4. *The 'Art of Individualisation.'* This invention of Tesla is to primitive 'tuning' what refined language is to unarticulated expression. It makes possible the transmission of signals or messages absolutely secret and exclusive both in the active and passive aspect, that is, non-interfering as well as non-interferable. Each signal is like an individual of unmistakable identity and there is virtually no limit to the number of stations or instruments which can be simultaneously operated without the slightest mutual disturbance.

"5. *The 'Terrestial Stationary Waves.'* This wonderful discovery, popularly explained, means that the Earth is responsive to electrical vibrations of definite pitch just as a tuning fork to certain waves of sound. These particular electrical vibrations, capable of powerfully exciting the Globe, lend themselves to innumerable uses of great importance commercially and in many other respects.

"The first 'World-System' power plant can be put in operation in nine months. With this power plant it will be practicable to attain electrical activities up to ten million horsepower and it is designed to serve for as many technical achievements as are possible without due expense. Among these the following may be mentioned:

"(1) The inter-connection of the existing telegraph exchanges or offices all over the world;

"(2) The establishment of a secret and non-interferable government telegraph service;

"(3) The inter-connection of all the present telephone exchanges or offices on the Globe;

"(4) The universal distribution of general news, by telegraph or telephone, in connection with the Press;

"(5) The establishment of such a 'World-System' of intelligence transmission for exclusive private use;

"(6) The inter-connection and operation of all stock tickers of the world;

"(7) The establishment of a 'World-System' of musical distribution, etc.;

"(8) The universal registration of time by cheap clocks indicating the hour with astronomical precision and requiring no attention whatever;

"(9) The world transmission of typed or handwritten characters, letters, checks, etc.;

"(10) The establishment of a universal marine service enabling the navigators of all ships to steer perfectly without compass, to determine the exact location, hour and speed, to prevent collisions and disasters, etc.;

"(11) The inauguration of a system of world-printing on land and sea;

"(12) The world reproduction of photographic pictures and all kinds of drawings or records."

I also proposed to make demonstrations in the wireless transmission of power on a small scale but sufficient to carry conviction. Besides these I referred to other and incomparably more important applications of my discoveries which will be disclosed at some future date.

A plant was built on Long Island with a tower 187 feet high, having a spherical terminal about 68 feet in diameter. These dimensions were adequate for the transmission of virtually any amount of energy. Originally only from 200 to 300 K.W. were provided but I intended to employ later several thousand horsepower. The transmitter was to emit a wave-complex of special characteristics and I had devised a unique method of telephonic control of any amount of energy.

The tower was destroyed two years ago but my projects are being developed and another one, improved in some features, will be constructed. On this occasion I would contradict the widely circulated report that the structure was demolished by the Government which owing to war conditions, might have created prejudice in the minds of those who may not know that the papers, which thirty years ago conferred upon me the honor of American citizenship, are always kept in a safe, while my orders, diplomas, degrees, gold medals and other distinctions are packed away in old trunks. If this report had a foundation I would have been refunded a large sum of money which I expended in the construction of the tower. On the contrary it was in the interest of the Government to preserve it, particularly as it would have made possible—to mention just one valuable result—the location of a submarine in any part of the world. My plant, services, and all my improvements have always been at the disposal of the officials and ever since the outbreak of the European conflict I have been working at a sacrifice on several inventions of mine relating to aerial navigation, ship propulsion and wireless transmission which are of the greatest importance to the country. Those who are well informed know that my ideas have revolutionized the industries of the United States and I am not aware that there lives an inventor who has been, in this respect, as fortunate as myself especially as regards the use of his improvements in the war. I have refrained from publicly expressing myself on this subject before as it seemed improper to dwell on personal matters while all the world was in dire trouble. I would add further, in view of various rumors which have reached me, that Mr. J. Pierpont Morgan did not interest himself with me in a business way but in the same large spirit in which he has assisted many other pioneers. He carried out his generous promise to the letter and it would have been most unreasonable to expect from him anything more. He had the highest regard for my attainments and gave me every evidence of his complete faith in my ability to ultimately achieve what I had set out to do. I am unwilling to accord to some small-minded and jealous individuals the satisfaction of having thwarted my efforts. These men are to me nothing more than microbes of a nasty disease. My project was retarded by laws of nature. The world was not prepared for it. It was too far ahead of time. But the same laws will prevail in the end and make it a triumphal success.

PART VI

THE ART OF
TELAUTOMATICS

How Tesla's Mind Recuperates

No subject to which I have ever devoted myself has called for such concentration of mind and strained to so dangerous a degree the finest fibers of my brain as the system of which the Magnifying Transmitter is the foundation. I put all the intensity and vigor of youth in the development of the rotating field discoveries, but those early labors were of a different character. Although strenuous in the extreme, they did not involve that keen and exhausting discernment which had to be exercised in attacking the many puzzling problems of the wireless. Despite my rare physical endurance at that period the abused nerves finally rebelled and I suffered a complete collapse, just as the consummation of the long and difficult task was almost in sight. Without doubt I would have paid a greater penalty later, and very likely my career would have been prematurely terminated, had not providence equipt me with a safety device, which has seemed to improve with advancing years and unfailingly comes into play when my forces are at an end. So long as it operates I am safe from danger, due to overwork, which threatens other inventors and, incidentally, I need no vacations which are indispensable to most people. When I am all but used up I simply do as the darkies, who "naturally fall asleep while white folks worry." To venture a theory out of my sphere—the body probably accumulates little by little a definite quantity of some toxic agent and I sink into a nearly lethargic state which lasts half an hour to the minute. Upon awakening I have the sensation as though the events immediately preceding had occurr very long ago, and if I attempt to continue the interrupted train of thought I feel a veritable mental nausea. Involuntarily I then turn to other work and am surprised at the freshness of the mind and ease with which I overcome obstacles that had baffled me before. After weeks or months my passion for the temporarily abandoned invention returns and I invariably find answers to all the vexing questions with scarcely any effort. In this connection I will tell of an extraordinary experience which may be of interest to students of psychology. I had produced a striking phenomenon with my grounded transmitter and was endeavoring to ascertain its true significance in relation to the currents propagated through the earth. It seemed a hopeless undertaking, and for more than a year I worked unremittingly, but in vain. This profound study so entirely absorbed me that I became forgetful of everything else, even of my undermined health. At last, as I was at the point of breaking down, nature applied the preservative inducing lethal sleep. Regaining my senses, I realized with consternation that I was unable to visualize scenes from my life except those of infancy, the very first ones that had entered my consciousness. Curiously enough, these appeared before my vision with startling distinctness and afforded me welcome relief. Night after night, when retiring, I would think of them and more and more of my previous existence was revealed. The image of my mother was always the principal figure in the spectacle that slowly unfolded, and a consuming desire to see her again gradually took possession of me. This feeling grew so strong that I resolved to drop all work and satisfy my longing. But I found it too hard to break away from the laboratory, and several months elapsed during which I had succeeded in reviving all the impressions of my past life up to the spring of 1892. In the next picture that came out of the mist of oblivion, I saw myself at the *Hotel de la Paix* in Paris just coming to from one of my peculiar sleeping spells, which had been caused by prolonged exertion of the brain. Imagine the pain and distress I felt when it flashed upon my mind that a dispatch was handed to me at that very moment bearing the sad news that my mother was dying; I remembered how I made the long journey home without an hour of rest and how she passed away after weeks of agony! It was especially remarkable that during all this period of partially obliterated memory I was fully alive to everything touching on the subject of my research. I could recall the smallest details and the least insignificant observations in my experiments and even recite pages of text and complex mathematical formulae.

My belief is firm in a law of compensation. The true rewards are ever in proportion to the labor and sacrifices made. This is one of the reasons why I feel certain that of all my inventions, the Magnifying Transmitter will prove most important and valuable to future generations. I am prompted to this prediction not so much by thoughts of the commercial and industrial revolution which it will surely bring about, but of the humanitarian consequences of the many achievements it makes possible. Considerations of mere utility weigh little in the balance against the higher benefits of civilization. We are confronted with portentous problems which can not be solved just by providing for our material existence, however abundantly. On the contrary, progress in this direction is fraught with hazards and perils not less menacing than those born from want and suffering. If we were to release the energy of atoms or discover some other way of developing cheap and unlimited power at any point of the globe this accomplishment, instead of being a blessing, might bring disaster to mankind in giving rise to dissension and anarchy which would ultimately result in the enthronement of the hated regime of force. The greatest good will comes from technical improvements tending to unification and harmony, and my wireless transmitter is preeminently such. By its means the human voice and likeness will be reproduced everywhere and factories driven thousands of miles from waterfalls furnishing the power; aerial machines will be propelled around the earth without a stop and the sun's energy controlled to create lakes and rivers for motive purposes and transformation of arid deserts into fertile land. Its introduction for telegraphic, telephonic and similar uses will automatically cut out the statics and all other interferences which at present impose narrow limits to the application of the wireless. This is a timely topic on which a few words might not be amiss.

Tesla Raps "Static" Men Vigorously

During the past decade a number of people have arrogantly claimed that they had succeeded in doing away with this impediment. I have carefully examined all of the arrangements described and tested most of them long before they were publicly disclosed, but the finding was uniformly negative. A recent official statement from the U. S. Navy may, perhaps, have taught some beguilable news editors how to appraise these announcements at their real worth. As a rule the attempts are based on theories so fallacious that whenever they come to my notice I can not help thinking in a lighter vein. Quite recently a new discovery was heralded, with a deafening flourish of trumpets, but it proved another case of a mountain bringing forth a mouse. This reminds me of an exciting incident which took place years ago when I was conducting my experiments with currents of high frequency. Steve Brodie had just jumped off the Brooklyn Bridge. The feat has been vulgarized since by imitators, but the first report electrified New York. I was very impressionable then and frequently spoke of the daring printer. On a hot afternoon I felt the necessity of refreshing myself and stepped into one of the popular thirty thousand institutions of this great city where a delicious twelve per cent beverage was served which can now be had only by making a trip to the poor and devastated countries of Europe. The attendance was large and not over-distinguished and a matter was discussed which gave me an admirable opening for the careless remark: "This is what I said when I jumped off the bridge." No sooner had I uttered these words than I felt like the companion of Timotheus in the poem of Schiller. In an instant there was a pandemonium and a dozen voices cried: "It is Brodie!" I threw a quarter on the counter and bolted for the door but the crowd was at my heels with yells: "Stop, Steve!" which must have been misunderstood for many persons tried to hold me up as I ran frantically for my haven of refuge. By darting around corners I fortunately managed—through the medium of a fire-escape—to reach the laboratory, where I threw off my coat, camouflaged myself as a hard-working blacksmith, and started the forge. But these precautions proved unnecessary; I had eluded my pursuers. For many years afterward, at night, when imagination turns into spectres the trifling troubles of the day, I often thought, as I tossed on the bed, what my fate would have been had that mob caught me and found out that I was not Steve Brodie!

Now the engineer, who lately gave an account before a technical body of a novel remedy against statics based on a "heretofore unknown law of nature," seems to have been as reckless as myself when he contended that these disturbances propagate up and down, while those of a transmitter proceed along the earth. It would mean that a condenser, as this globe, with its gaseous envelope, could be charged and discharged in a manner quite contrary to the fundamental teachings propounded in every elemental text-book of physics. Such a supposition would have been condemned as erroneous, even in Franklin's time, for the facts bearing on this were then well known and the identity between atmospheric electricity and that developed by machines was fully established. Obviously, natural and artificial disturbances propagate through the earth and the air in exactly the same way, and both set up electro-motive forces in the horizontal, as well as vertical, sense. Interference can not be overcome by any such methods as were proposed. The truth is this: In the air the potential increases at the rate of about fifty volts per foot of elevation, owing to which there may be a difference of pressure amounting to twenty, or even forty thousand volts between the upper and lower ends of the antenna. The masses of the charged atmosphere are constantly in motion and give up electricity to the conductor, not continuously but rather disruptively, this producing a grinding noise in a sensitive telephonic receiver. The higher the terminal and the greater the space encompast by the wires, the more pronounced is the effect, but it must be understood that it is purely local and has little to do with the real trouble. In 1900, while perfecting my wireless system, one form of apparatus comprised four antennae. These were carefully calibrated to the same frequency and connected in multiple with the object of magnifying the action, in receiving from any direction. When I desired to ascertain the origin of the transmitted impulses, each diagonally situated pair was put in series with a primary coil energizing the detector circuit. In the former case the sound was loud in the telephone; in the latter it ceased, as expected, the two antennae neutralizing each other, but the true statics manifested themselves in both instances and I had to devise special preventives embodying different principles.

The Remedy for Static

By employing receivers connected to two points of the ground, as suggested by me long ago, this trouble caused by the charged air, which is very serious in the structures as now built, is nullified and besides, the liability of all kinds of interference is reduced to about one-half, because of the directional character of the circuit. This was perfectly self-evident, but came as a revelation to some simple-minded wireless folks whose experience was confined to forms of apparatus that could have been improved with an axe, and they have been disposing of the bear's skin before killing him. If it were true that strays performed such antics, it would be easy to get rid of them by receiving without aerials. But, as a matter of fact, a wire buried in the ground which, conforming to this view, should be absolutely immune, is more susceptible to certain extraneous impulses than one placed vertically in the air. To state it fairly, a slight progress has been made, but not by virtue of any particular method or device. It was achieved simply by discarding the enormous structures, which are bad enough for transmission but wholly unsuitable for reception, and adopting a more appropriate type of receiver. As I pointed out in a previous article, to dispose of this difficulty for good, a radical change must be made in the system, and the sooner this is done the better.

Radio Government Control
Not Wanted

It would be calamitous, indeed, if at this time when the art is in its infancy and the vast majority, not excepting even experts, have no conception of its ultimate possibilities, a measure would be rushed through the legislature making it a government monopoly. This was proposed a few weeks ago by Secretary Daniels, and no doubt that distinguished official has made his appeal to the Senate and House of Representatives with sincere conviction. But universal evidence unmistakably shows that the best results are always obtained in healthful commercial competition. There are, however, exceptional reasons why wireless should be given the fullest freedom of development. In the first place it offers prospects immeasurably greater and more vital to betterment of human life than any other invention or discovery in the history of man. Then again, it must be understood that this wonderful art has been, in its entirety, evolved here and can be called "American" with more right and propriety than the telephone, the incandescent lamp or the aeroplane. Enterprising press agents and stock jobbers have been so successful in spreading misinformation that even so excellent a periodical as the *Scientific American* accords the chief credit to a foreign country. The Germans, of course, gave us the Hertz-waves and the Russian, English, French and Italian experts were quick in using them for signaling purposes. It was an obvious application of the new agent and accomplished with the old classical and unimproved induction coil—scarcely anything more than another kind of heliography. The radius of transmission was very limited, the results attained of little value, and the Hertz oscillations, as a means for conveying intelligence, could have been advantageously replaced by sound-waves, which I advocated in 1891. Moreover, all of these attempts were made three years after the basic principles of the wireless system, which is universally employed to-day, and its potent instrumentalities had been clearly described and developed in America. No trace of those Hertzian appliances and methods remains today. We have proceeded in the very opposite direction and what has been done is the product of the brains and efforts of citizens of this country. The fundamental patents have expired and the opportunities are open to all. The chief argument of the Secretary is based on interference. According to his statement, reported in the New York *Herald* of July 29th, signals from a powerful station can be intercepted in every village of the world. In view of this fact, which was demonstrated in my experiments of 1900, it would be of little use to impose restrictions in the United States.

America First

As throwing light on this point, I may mention that only recently an odd looking gentleman called on me with the object of enlisting my services in the construction of world transmitters in some distant land. "We have no money," he said, "but carloads of solid gold and we will give you a liberal amount." I told him that I wanted to see first what will be done with my inventions in America, and this ended the interview. But I am satisfied that some dark forces are at work, and as time goes on the maintenance of continuous communication will be rendered more difficult. The only remedy is a system immune against interruption. It has been perfected, it exists, and all that is necessary is to put it in operation.

The terrible conflict is still uppermost in the minds and perhaps the greatest importance will be attached to the Magnifying Transmitter as a machine for attack and defense, more particularly in connection with *Telautomatics*. This invention is a logical outcome of observations begun in my boyhood and continued thruout my life. When the first results were publisht the *Electrical Review* stated editorially that it would become one of the "most potent factors in the advance and civilization of mankind." The time is not distant when this prediction will be fulfilled. In 1898 and 1900 it was offered to the Government and might have been adopted were I one of those who would go to Alexander's shepherd when they want a favor from Alexander. At that time I really thought that it would abolish war, because of its unlimited destructiveness and exclusion of the personal element of combat. But while I have not lost faith in its potentialities, my views have changed since.

The Road to Permanent Peace

War can not be avoided until the physical cause for its recurrence is removed and this, in the last analysis, is the vast extent of the planet on which we live. Only thru annihilation of distance in every respect as, the conveyance of intelligence, transport of passengers and supplies and transmission of energy will conditions be brought about some day, insuring permanency of friendly relations. What we now want most is closer contact and better understanding between individuals and communities all over the earth, and the elimination of that fanatic devotion to exalted ideals of national egoism and pride which is always prone to plunge the world into primeval barbarism and strife. No league or parliamentary act of any kind will ever prevent such a calamity. These are only new devices for putting the weak at the mercy of the strong. I have exprest myself in this regard fourteen years ago, when a combination of a few leading governments—a sort of Holy Alliance—was advocated by the late Andrew Carnegie, who may be fairly considered as the father of this idea, having given to it more publicity and impetus than anybody else prior to the efforts of the President. While it can not be denied that such a pact might be of material advantage to some less fortunate peoples, it can not attain the chief object sought. Peace can only come as a natural consequence of universal enlightenment and merging of races, and we are still far from this blissful realization. As I view the world of today, in the light of the gigantic struggle we have witnest, I am filled with conviction that the interests of humanity would be best served if the United States remained true to its traditions and kept out of "entangling alliances." Situated as it is, geographically, remote from the theaters of impending conflicts, without incentive to territorial aggrandizement, with inexhaustible resources and immense population thoroly imbued with the spirit of liberty and right, this country is placed in a unique and privileged position. It is thus able to exert, independently, its colossal strength and moral force to the benefit of all, more judiciously and effectively, than as member of a league.

The Mechanistic Theory of Life

In one of these biographical sketches, published in the ELECTRICAL EXPERIMENTER, I have dwelt on the circumstances of my early life and told of an affliction which compelled me to unremitting exercise of imagination and self-observation. This mental activity, at first involuntary under the pressure of illness and suffering, gradually became second nature and led me finally to recognize that I was but an automaton devoid of free will in thought and action and merely responsive to the forces of the environment. Our bodies are of such complexity of structure, the motions we perform are so numerous and involved, and the external impressions on our sense organs to such a degree delicate and elusive that it is hard for the average person to grasp this fact. And yet nothing is more convincing to the trained investigator than the mechanistic theory of life which had been, in a measure, understood and propounded by Descartes three hundred years ago. But in his time many important functions of our organism were unknown and, especially with respect to the nature of light and the construction and operation of the eye, philosophers were in the dark. In recent years the progress of scientific research in these fields has been such as to leave no room for a doubt in regard to this view on which many works have been published. One of its ablest and most eloquent exponents is, perhaps, Felix Le Dantoc, formerly assistant of Pasteur. Prof. Jacques Loeb has performed remarkable experiments in heliotropism, clearly establishing the controlling power of light in lower forms of organisms, and his latest book, "Forced Movements," is revelatory. But while men of science accept this theory simply as any other that is recognized, to me it is a truth which I hourly demonstrate by every act and thought of mine. The consciousness of the external impression prompting me to any kind of exertion, physical or mental, is ever present in my mind. Only on very rare occasions, when I was in a state of exceptional concentration, have I found difficulty in locating the original impulses.

Lack of Observation a Form of Ignorance

The by far greater number of human beings are never aware of what is passing around and within them, and millions fall victims of disease and die prematurely just on this account. The commonest, every-day occurrences appear to them mysterious and inexplicable. One may feel a sudden wave of sadness and rake his brain for an explanation when he might have noticed that it was caused by a cloud cutting off the rays of the sun. He may see the image of a friend dear to him under conditions which he construes as very peculiar, when only shortly before he has passed him in the street or seen his photograph somewhere. When he loses a collar button he fusses and swears for an hour, being unable to visualize his previous actions and locate the object directly. Deficient observation is merely a form of ignorance and responsible for the many morbid notions and foolish ideas prevailing. There is not more than one out of every ten persons who does not believe in telepathy and other psychic manifestations, spiritualism and communion with the dead, and who would refuse to listen to willing or unwilling deceivers. Just to illustrate how deeply rooted this tendency has become even among the clear-headed American population, I may mention a comical incident.

Psychic Phenomena in
the Manufacture of Flivvers

Shortly before the war, when the exhibition of my turbines in this city elicited widespread comment in the technical papers, I anticipated that there would be a scramble among manufacturers to get hold of the invention, and I had particular designs on that man from Detroit who has an uncanny faculty for accumulating millions. So confident was I that he would turn up some day, that I declared this as certain to my secretary and assistants. Sure enough, one fine morning a body of engineers from the Ford Motor Company presented themselves with the request of discussing with me an important project. "Didn't I tell you?" I remarked triumphantly to my employees, and one of them said, "You are amazing, Mr. Tesla; everything comes out exactly as you predict." As soon as these hard-headed men were seated I, of course, immediately began to extol the wonderful features of my turbine, when the spokesmen interrupted me and said, "We know all about this, but we are on a special errand. We have formed a psychological society for the investigation of psychic phenomena and we want you to join us in this undertaking." I suppose those engineers never knew how near they came to being fired out of my office.

Confuting Spiritism

Ever since I was told by some of the greatest men of the time, leaders in science whose names are immortal, that I am possesst of an unusual mind, I bent all my thinking faculties on the solution of great problems regardless of sacrifice. For many years I endeavored to solve the enigma of death, and watched eagerly for every kind of spiritual indication. But only once in the course of my existence have I had an experience which momentarily impressed me as supernatural. It was at the time of my mother's death. I had become completely exhausted by pain and long vigilance, and one night was carried to a building about two blocks from our home. As I lay helpless there,

I thought that if my mother died while I was away from her bedside she would surely give me a sign. Two or three months before I was in London in company with my late friend, Sir William Crookes, when spiritualism was discussed, and I was under the full sway of these thoughts. I might not have paid attention to other men, but was susceptible to his arguments as it was his epochal work on radiant matter, which I had read as a student, that made me embrace the electrical career. I reflected that the conditions for a look into the beyond were most favorable, for my mother was a woman of genius and particularly excelling in the powers of intuition. During the whole night every fiber in my brain was strained in expectancy, but nothing happened until early in the morning, when I fell in a sleep, or perhaps a swoon, and saw a cloud carrying angelic figures of marvelous beauty, one of whom gazed upon me lovingly and gradually assumed the features of my mother. The appearance slowly floated across the room and vanished, and I was awakened by an indescribably sweet song of many voices. In that instant a certitude, which no words can express, came upon me that my mother had just died. And that was true. I was unable to understand the tremendous weight of the painful knowledge I received in advance, and wrote a letter to Sir William Crookes while still under the domination of these impressions and in poor bodily health. When I recovered I sought for a long time the external cause of this strange manifestation and, to my great relief, I succeeded after many months of fruitless effort. I had seen the painting of a celebrated artist, representing allegorically one of the seasons in the form of a cloud with a group of angels which seemed to actually float in the air, and this had struck me forcefully. It was exactly the same that appeared in my dream, with the exception of my mother's likeness. The music came from the choir in the church nearby at the early mass of Easter morning, explaining everything satisfactorily in conformity with scientific facts.

This occurred long ago, and I have never had the faintest reason since to change my views on psychical and spiritual phenomena, for which there is absolutely no foundation. The belief in these is the natural outgrowth of intellectual development.

Religious dogmas are no longer accepted in their orthodox meaning, but every individual clings to faith in a supreme power of some kind. We all must have an ideal to govern our conduct and insure contentment, but it is immaterial whether it be one of creed, art, science or anything else, so long as it fulfills the function of a dematerializing force. It is essential to the peaceful existence of humanity as a whole that one common conception should prevail.

Tesla's Astounding Discovery

While I have failed to obtain any evidence in support of the contentions of psychologists and spiritualists, I have proved to my complete satisfaction the automatism of life, not only through continuous observations of individual actions, but even more conclusively through certain generalizations. These amount to a discovery which I consider of the greatest moment to human society, and on which I shall briefly dwell. I got the first inkling of this astounding truth when I was still a very young man, but for many years I interpreted what I noted simply as coincidences. Namely, whenever either myself or a person to whom I was attached, or a cause to which I was devoted, was hurt by others in a particular way, which might be best popularly characterized as the most unfair imaginable, I experienced a singular and undefinable pain which, for want of a better term, I have qualified as "cosmic," and shortly thereafter, and invariably, those who had inflicted it came to grief. After many such cases I confided this to a number of friends, who had the opportunity to convince themselves of the truth of the theory which I have gradually formulated and which may be stated in the following few words:

Our bodies are of similar construction and exposed to the same external influences. This results in likeness of response and concordance of the general activities on which all our social and other rules and laws are based. We are automata entirely controlled by the forces of the medium, being tossed about like corks on the surface of the water, but mistaking the resultant of the impulses from the outside for free will. The movements and other actions we perform are always life preservative and tho seemingly quite independent from one another, we are connected by invisible links. So long as the organism is in perfect order it responds accurately to the agents that prompt it, but the moment that there is some derangement in any individual, his self-preservative power is impaired. Everybody understands, of course, that if one becomes deaf, has his eyesight weakened, or his limbs injured, the chances for his continued existence are lessened. But this is also true, and perhaps more so, of certain defects in the brain which deprive the automaton, more or less, of that vital quality and cause it to rush into destruction. A very sensitive and observant being, with his highly developed mechanism all intact, and acting with precision in obedience to the changing conditions of the environment, is endowed with a transcending mechanical sense, enabling him to evade perils too subtle to be directly perceived. When he comes in contact with others whose controlling organs are radically faulty, that sense asserts itself and he feels the "cosmic" pain. The truth of this has been borne out in hundreds of instances and I am inviting other students of nature to devote attention to this subject, believing that thru combined and systematic effort results of incalculable value to the world will be attained.

Dr. Tesla's First Automaton

The idea of constructing an automaton, to bear out my theory, presented itself to me early but I did not begin active work until 1893, when I started my wireless investigations. During the succeeding two or three years a number of automatic mechanisms, to be actuated from a distance, were constructed by me and exhibited to visitors in my laboratory. In 1896, however, I designed a complete machine capable of a multitude of operations, but the consummation of my labors was delayed until late in 1897. This machine was illustrated and described in my article in the Century Magazine of June, 1900, and other periodicals of that time and, when first shown in the beginning of 1898, it created a sensation such as no other invention of mine has ever produced. In November, 1898, a basic patent on the novel art was granted to me, but only after the Examiner-in-Chief had come to New York and witnesst the performance, for what I claimed seemed unbelievable. I remember that when later I called on an official in Washington, with a view of offering the invention to the Government, he burst out in laughter upon my telling him what I had accomplished. Nobody thought then that there was the faintest prospect of perfecting such a device. It is unfortunate that in this patent, following the advice of my attorneys, I indicated the control as being effected thru the medium of a single circuit and a well-known form of detector, for the reason that I had not yet secured protection on my methods and apparatus for individualization. As a matter of fact, my boats were controlled thru the joint action of several circuits and interference of every kind was excluded. Most generally I employed receiving circuits in the form of loops, including condensers, because the discharges of my high-tension transmitter ionized the air in the hall so that even a very small aerial would draw electricity from the surrounding atmosphere for hours. Just to give an idea, I found, for instance, that a bulb 12" in diameter, highly exhausted, and with one single terminal to which a short wire was attached, would deliver well on to one thousand successive flashes before all charge of the air in the laboratory was neutralized. The loop form of receiver was not sensitive to such a disturbance and it is curious to note that it is becoming popular at this late date. In reality it collects much less energy than the aerials or a long grounded wire, but it so happens that it does away with a number of defects inherent to the present wireless devices. In demonstrating my invention before audiences, the visitors were requested to ask any questions, however involved, and the automaton would answer them by signs. This was considered magic at that time but was extremely simple, for it was myself who gave the replies by means of the device.

At the same period another larger telautomatic boat was constructed, a photograph of which is shown in this number of the ELECTRICAL EXPERIMENTER. It was controlled by loops, having several turns placed in the hull, which was made entirely water-tight and capable of submergence. The apparatus was similar to that used in the first with the exception of certain special features I introduced as, for example, incandescent lamps which afforded a visible evidence of the proper functioning of the machine.

Telautomatics of the Future

These automata, controlled within the range of vision of the operator, were, however, the first and rather crude steps in the evolution of the Art of Telautomatics as I had conceived it. The next logical improvement was its application to automatic mechanisms beyond the limits of vision and at great distance from the center of control, and I have ever since advocated their employment as instruments of warfare in preference to guns. The importance of this now seems to be recognized, if I am to judge from casual announcements thru the press of achievements which are said to be extraordinary but contain no merit of novelty, whatever. In an imperfect manner it is practicable, with the existing wireless plants, to launch an aeroplane, have it follow a certain approximate course, and perform some operation at a distance of many hundreds of miles. A machine of this kind can also be mechanically controlled in several ways and I have no doubt that it may prove of some usefulness in war. But there are, to my best knowledge, no instrumentalities in existence today with which such an object could be accomplished in a precise manner. I have devoted years of study to this matter and have evolved means, making such and greater wonders easily realizable. As stated on a previous occasion, when I was a student at college I conceived a flying machine quite unlike the present ones. The underlying principle was sound but could not be carried into practice for want of a prime-mover of sufficiently great activity. In recent years I have successfully solved this problem and am now planning aerial machines devoid of sustaining planes, ailerons, propellers and other external attachments, which will be capable of immense speeds and are very likely to furnish powerful arguments for peace in the near future. Such a machine, sustained and propelled entirely by reaction, is shown on one of the pages and is supposed to be controlled either mechanically or by wireless energy. By installing proper plants it will be practicable to project a missile of this kind into the air and drop it almost on the very spot designated, which may be thousands of miles away. But we are not going to stop at this. Telautomata will be ultimately produced, capable of acting as if possesst of their own intelligence, and their advent will create a revolution. As early as 1898 I proposed to representatives of a large manufacturing concern the construction and public exhibition of an automobile carriage which, left to itself, would perform a great variety of operations involving something akin to judgment. But my proposal was deemed chimerical at that time and nothing came from it.

At present many of the ablest minds are trying to devise expedients for preventing a repetition of the awful conflict which is only theoretically ended and the duration and main issues of which I have correctly predicted in an article printed in the *Sun* of December 20, 1914. The proposed League is not a remedy but, on the contrary, in the opinion of a number of competent men, may bring about results just the opposite. It is particularly regrettable that a punitive policy was adopted in framing the terms of peace, because a few years hence it will be possible for nations to fight without armies, ships or guns, by weapons far more terrible, to the destructive action and range of which there is virtually no limit. Any city, at a distance, whatsoever, from the enemy, can be destroyed by him and no power on earth can stop him from doing so. If we want to avert an impending calamity and a state of things which may transform this globe into an inferno, we should push the development of flying machines and wireless transmission of energy without an instant's delay and with all the power and resources of the nation.

Other Writings

TESLA WOULD POUR LIGHTNING FROM AIRSHIPS TO CONSUME FOE

By Dr. Nikola Tesla

*Inventor of Wireless Power Transmission
and Discoverer of the High Frequency,
Alternating Electrical Current*

(In an Interview.)

At this moment there is available to the belligerents in Europe an effective means by which the present conflict could be brought to an abrupt termination within 24 hours—and which would forever after make war an impossible thing.

The various parliaments for peace which have been initiated are laudable notions. But they are only the shadows of hope; dreams built upon a beautiful if not an altogether impracticable philosophy.

With immeasurably less ado, with certainty and dispatch the men may be removed from the trenches; arbitrarily, unconditionally and probably without loss of life—not ever again to return.

War Is a Thousand Infernos

The means proposed to accomplish this would immediately and automatically materialize reality by the application of an existent, concrete force—the brutal fact versus abstract theory.

So soon as either the entente allies or the central powers should develop and adopt the invention described the war would end. To continue would result in the sure and speedy annihilation of the nations against which it might be employed.

Human resistance to this terrific instrument of death and devastation would be physically impossible. It could not be checked or combated by any means now in existence. And it would be as irresistible as it would be effective.

War is without reason or excuse; a horror greater than a thousand infernos, real or Dante-pictured. Engines of slaughter have been the creations of a war-mad people. They are symbols of a barbaric blood lust—without the pale of civilization and Christianity.

Yet we have wars. And as a natural accompaniment there are employed the most improved death-dealing machines of the age.

Men Mere Pawns Now

In ancient times men fought man to man, with the strength and sinews of their arms, with skill and might of sword and lance, hand to hand. And prowess and brawn and stamina and mastership—individual supremacy in generalship—won their battles, much as they are won today in the boxing ring. In modern warfare men are merely pawns, their fighting ability supplanted by the destroying capacity of machinery. And there have recently been produced such tremendous and before undreamed-of instruments of destruction that the world stands paralyzed with amazement and horror at the havoc they work.

However, even though it seem a paradox, it remained for the nation most disposed to peace to create a yet greater agency of destruction than any thus far contrived. Not one which merely will surpass present weapons in degree, but one which will out-Herod Herod in battle equipment, one which will prove an all-efficient means of widespread annihilation on a gigantic scale; against which no army or nation could survive, and against which there could be offered no adequate defense or protection.

Either Side Could End It

And it is now possible for either of the principals at war readily and quickly to devise and construct an effective equipment of such destructive apparatus in sufficient numbers to bring the enemy nation to cry for peace. It would make little or no difference as to the fact of terminating hostilities if both factions should equally so equip themselves at the same time, since the equipment to which I refer could not effectively be employed either in offensive or defensive operations against similar equipment. The nation first perfecting and adopting it would have the advantage in that they would be in a position to name the terms of settlement.

Practically no new thing in the arts and sciences would need to be invented or created for the production of such an engine of destruction. The various fundamental elements and essential parts required in its construction already are developed and generally in use. It remains only for the designer, the artisan, the electrician and the manufacturer to cooperate in producing and assembling them into a composite whole.

Employ Aircraft for Demolition

It may be conceded that the bomb air raids by dirigibles thus far have failed of decisive results; that they have only added to the general suffering and horror of it all. And—here is the point:

These same air craft might readily be converted into means by which cities and forts and munitions factories could be reduced to beds of whitened ashes and which could exterminate every living thing from the face of the earth.

One single demonstration of the stupendous demolition made possible by such a weapon would be all sufficient to bring the opposing forces instantly to sue for terms of peace.

Although several devices might be employed for this purpose, I shall cite but one to show how readily either enemy faction could establish a method of warfare which would make for the immediate termination of war.

A Vast Incendiary Arc

An airship of the gas or Zeppelin type developing 200 to 400 horsepower could be equipped with a high voltage electrical generator. For the project intended almost the full energy of the engines might be devoted to driving the dynamo, since the speed of the craft in operation would be a secondary consideration.

From a suitable reel in electrical connection with the dynamo thin wires would be lowered until their nether ends reached to within a certain distance from the ground or the tops of the buildings of a city. Through these suspended wires there would be discharged several million volt currents of electricity.

Between the lower ends of the wires and the earth there would then be formed a tremendous arc, which would consume with its terrific heat not only every human being and living thing, but also every building and every atom of combustible material over and near which this incendiary arc should pass.

The destructive radius would be at least 75 to 100 yards from the electrical generation of engines of, say, 200-horsepower, although with engines of 500-horsepower the radius of incendiary effect might easily be 200 to 300 yards, or even more; especially if the wire area were extended by the use of stretcher bars under the airship.

A highly sensitive automatic regulator controlling the reels constantly would adjust the length of the lowered wires to maintain a proper distance between the discharge ends of the wire and the earth to produce and preserve an uninterrupted arc.

If the craft were provided with auxiliary engines for motive power, it might travel with considerable speed, sweeping a path clean of every form of life and every vestige of man's creation. Also, such a craft could sail at a great height—sufficiently high, in fact, to be barely visible to the naked human eye, and certainly well beyond the range of antiaircraft guns.

Surrender Swift and Final

Two or three such arcships side by side traveling parallel with an enemy's lines would leave in their wake a mass of smoldering ruins of bomb-proof trenches, supply trains. munition wagons, gun carriages, fences, homesteads and villages, and strewn through all the charred and fire-shriveled bodies of soldiers fallen helpless before this irresistible and all-consuming force.

Particularly if such an arcship raid were made at night the chances of repulsing or even reaching the enemy fleet would be practically nil.

No sane government would hesitate about flying the white flag in face of such a grewsome cataclysm. Surrender, swift and final, would be the only human alternative.

Burning Helpless Women

Moving in squadrons or line abreast, a fleet of these aircraft could sweep over large cities, even such as London, Berlin, Paris, Vienna, Rome, Petrograd or Constantinople, and leave them a blackened, smoking and lifeless ruin.

Perhaps not even the most war-crazed power ever would venture such an immeasurable and unpardonable outrage against mankind, involving as it would the burning of helpless women and children and noncombatants.

And yet such heinous offenses against civilization and the once recognized rules of war have been and still are being committed. Even though it be on an insignificant scale as compared with the possibilities of the device suggested for the arc craft, yet one may not forget that such incendiary raids are being carried on to the limit of the means thus far developed.

What they might attempt with more destructive agencies is seriously open to question.

The Humanitarian Side

However, it is largely due to the fact that the aerial raids with
present equipment do not make for decisive results that the
powers continue to tolerate such revolting tactics of warfare,
electing to take the gambler's chance of escape in future.

But—and this is the humanitarian side to this suggested
engine of annihilation—if either belligerent should equip and
place in active service an effective fleet of these arc-producing,
incendiary air craft and should make due announcement to
enemy nations of a proposed raid upon its cities or its forces in
the field, such an act would be equivalent to serving notice of a
demand for surrender—and beyond peradventure it would
promptly and unconditionally be accepted.

THE ACTION OF
THE EYE

It can be taken as a fact, which the theory of the action of the eye implies, that, for each external impression, that is, for each image produced on the retina, the ends of the visual nerves concerned in the conveyance of the impression to the mind, must be under a peculiar stress or in a vibratory state. It now does not seem improbable that, when by the power of thought an image is evoked, a distant reflex action, no matter how weak, is exerted upon certain ends of the visual nerves, and, therefore, upon the retina. Will it ever be within human power to analyze the condition of the retina when disturbed by thought or reflex action, by the help of some optical or other means of such sensitiveness that a clear idea of its state might be gained at any time? If this were possible, then the problem of reading one's thoughts with precision, like the characters of an open book, might be much easier to solve than many problems belonging to the domain of positive physical science, in the solution of which many, if not the majority, of scientific men implicitly believe. Helmholtz has shown that the fundi of the eyes are themselves luminous, and he was able to *see*, in total darkness, the movement of his arm by the light of his own eyes. This is one of the most remarkable experiments recorded in the history of science, and probably only a few men could satisfactorily repeat it, for it is very likely that the luminosity of the eyes is associated with uncommon activity of the brain and great imaginative power. It is fluorescence of brain action, as it were.

Another fact having a bearing on this subject which has probably been noted by many, since it is stated in popular expressions, but which I cannot recollect to have found chronicled as a positive result of observation, is, that at times, when a sudden idea or image presents itself to the intellect, there is a distinct and sometimes painful sensation of luminosity produced in the eye, observable even in broad day-light.

Two facts about the eye must forcibly impress the mind of the physicist, notwithstanding he may think or say that it is an imperfect optical instrument, forgetting that the very conception of that which is perfect or seems so to him, has been gained through this same instrument. Firstly, the eye is, as far as our positive knowledge goes, the only organ which is *directly* affected by that subtle medium, which, as science teaches us, must fill all space; secondly, it is the most sensitive of our organs, incomparably more sensitive to external impressions than any other.

This divine organ of sight, this indispensable instrument for thought and all intellectual enjoyment, which lays open to us the marvels of this universe, through which we have acquired what knowledge we possess, and which prompts us to and controls all our physical and mental activity,—by what is it affected? By light! What is light?

It is beyond the scope of my lecture to dwell upon the subject of light in general, my object being merely to bring presently to your notice a certain class of light effects and a number of phenomena observed in pursuing the study of these effects. But to be consistent in my remarks it is necessary to state that according to the idea now accepted by the majority of scientific men as a positive result of theoretical and experimental investigation, the various forms of manifestation of energy which were generally designated as "electric," or more precisely "electro-magnetic," are energy manifestations of the same nature as those of radiant heat and light. Therefore the phenomena of light and heat, and others besides these, may be called electrical phenomena. Thus electrical science has become the mother science of all, and its study has become all-important. The day when we shall know exactly what "electricity" is, will chronicle an event probably greater and more important than any other recorded in the history of the human race.

THE PROBLEM OF
INCREASING
HUMAN ENERGY

With Special Reference to the
Harnessing of the Sun's Energy

The Onward Movement of Man—
The Energy of the Movement—
The Three Ways of Increasing
Human Energy

Of all the endless variety of phenomena which nature presents to our senses, there is none that fills our minds with greater wonder than that inconceivably complex movement which, in its entirety, we designate as human life. Its mysterious origin is veiled in the forever impenetrable mist of the past, its character is rendered incomprehensible by its infinite intricacy, and its destination is hidden in the unfathomable depths of the future. Whence does it come? What is it? Whither does it tend? are the great questions which the sages of all times have endeavored to answer.

Modern science says: The sun is the past, the earth is the present, the moon is the future. From an incandescent mass we have originated, and into a frozen mass we shall turn. Merciless is the law of nature, and rapidly and irresistibly we are drawn to our doom. Lord Kelvin, in his profound meditations, allows us only a short span of life, something like six million years, after which time the sun's bright light will have ceased to shine, and its life-giving heat will have ebbed away, and our own earth will be a lump of ice, hurrying on through the eternal night. But do not let us despair. There will still be left on it a glimmering spark of life, and there will be a chance to kindle a new fire on some distant star. This wonderful possibility seems, indeed, to exist, judging from Professor Dewar's beautiful experiments with liquid air, which show that germs of organic life are not destroyed by cold, no matter how intense;

consequently they may be transmitted through the interstellar space. Meanwhile the cheering lights of science and art, ever increasing in intensity, illuminate our path, and the marvels they disclose, and the enjoyments they offer, make us measurably forgetful of the gloomy future.

Though we may never be able to comprehend human life, we know certainly that it is a movement, of whatever nature it be. The existence of a movement unavoidably implies a body which is being moved and a force which is moving it. Hence, wherever there is life, there is a mass moved by a force. All mass possesses inertia, all force tends to persist. Owing to this universal property and condition, a body, be it at rest or in motion, tends to remain in the same state, and a force, manifesting itself anywhere and through whatever cause, produces an equivalent opposing force, and as an absolute necessity of this it follows that every movement in nature must be rhythmical. Long ago this simple truth was clearly pointed out by Herbert Spencer, who arrived at it through a somewhat different process of reasoning. It is borne out in everything we perceive—in the movement of a planet, in the surging and ebbing of the tide, in the reverberations of the air, the swinging of a pendulum, the oscillations of an electric current, and in the infinitely varied phenomena of organic life. Does not the whole of human life attest it? Birth, growth, old age, and death of an individual, family, race, or nation, what is it all but a rhythm? All life-manifestation, then, even in its most intricate form, as exemplified in man, however involved and inscrutable, is only a movement, to which the same general laws of movement which govern throughout the physical universe must be applicable.

When we speak of man, we have a conception of humanity as a whole, and before applying scientific methods to the investigation of his movement, we must accept this as a physical fact. But can any one doubt to-day that all the millions of individuals and all the innumerable types and characters constitute an entity, a unit? Though free to think and act, we are held together, like the stars in the firmament, with ties inseparable. These ties we cannot see, but we can feel them. I cut myself in the finger, and it pains me: this finger is a part of me.

I see a friend hurt, and it hurts me, too: my friend and I are one. And now I see stricken down an enemy, a lump of matter which, of all the lumps of matter in the universe, I care least for, and still it grieves me. Does this not prove that each of us is only a part of a whole?

For ages this idea has been proclaimed in the consummately wise teachings of religion, probably not alone as a means of insuring peace and harmony among men, but as a deeply founded truth. The Buddhist expresses it in one way, the Christian in another, but both say the same: We are all one. Metaphysical proofs are, however, not the only ones which we are able to bring forth in support of this idea. Science, too, recognizes this connectedness of separate individuals, though not quite in the same sense as it admits that the suns, planets, and moons of a constellation are one body, and there can be no doubt that it will be experimentally confirmed in times to come, when our means and methods for investigating psychical and other states and phenomena shall have been brought to great perfection. Still more: this one human being lives on and on. The individual is ephemeral, races and nations come and pass away, but man remains. Therein lies the profound difference between the individual and the whole. Therein, too, is to be found the partial explanation of many of those marvelous phenomena of heredity which are the result of countless centuries of feeble but persistent influence.

Conceive, then, man as a mass urged on by a force. Though this movement is not of a translatory character, implying change of place, yet the general laws of mechanical movement are applicable to it, and the energy associated with this mass can be measured, in accordance with well-known principles, by half the product of the mass with the square of a certain velocity. So, for instance, a cannon-ball which is at rest possesses a certain amount of energy in the form of heat, which we measure in a similar way. We imagine the ball to consist of innumerable minute particles, called atoms or molecules, which vibrate or whirl around one another. We determine their masses and velocities, and from them the energy of each of these minute systems, and adding them all together, we get an idea of the total heat-energy contained in the ball, which is only seem-ingly at rest. In this purely theoretical estimate this energy may then be calculated by multiplying half of the total mass—that is, half of the sum of all the small masses—with the square of a velocity which is determined from the velocities of the sep-arate particles. In like manner we may conceive of human energy being measured by half the human mass multiplied with the square of a velocity which we are not yet able to compute. But our deficiency in this knowledge will not vitiate the truth of the deductions I shall draw, which rest on the firm basis that the same laws of mass and force govern throughout nature.

Man, however, is not an ordinary mass, consisting of spin-ning atoms and molecules, and containing merely heat-energy. He is a mass possessed of certain higher qualities by reason of the creative principle of life with which he is endowed. His mass, as the water in an ocean wave, is being continuously exchanged, new taking the place of the old. Not only this, but he grows, propagates, and dies, thus altering his mass inde-pendently, both in bulk and density. What is most wonderful of all, he is capable of increasing or diminishing his velocity of movement by the mysterious power he possesses of appropri-ating more or less energy from other substance, and turning it into motive energy. But in any given moment we may ignore these slow changes and assume that human energy is meas-ured by half the product of man's mass with the square of a certain hypothetical velocity. However we may compute this velocity, and whatever we may take as the standard of its meas-ure, we must, in harmony with this conception, come to the conclusion that the great problem of science is, and always will be, to increase the energy thus defined. Many years ago, stim-ulated by the perusal of that deeply interesting work, Draper's "History of the Intellectual Development of Europe," depict-ing so vividly human movement, I recognized that to solve this eternal problem must ever be the chief task of the man of sci-ence. Some results of my own efforts to this end I shall endeavor briefly to describe here.

Let, then, in diagram *a*, M represent the mass of man. This mass is impelled in one direction by a force *f*, which is resisted by another partly frictional and partly negative force R, acting in a direction exactly opposite, and retarding the movement of the mass. Such an antagonistic force is present in every movement, and must be taken into consideration. The difference between these two forces is the effective force which imparts a velocity V to the mass M in the direction of the arrow on the line representing the force f. In accordance with the preceding, the human energy will then be given by the product $\frac{1}{2}MV^2 = \frac{1}{2}MV \times V$, in which M is the total mass of man in the ordinary interpretation of the term "mass," and V is a certain hypothetical velocity, which, in the present state of science, we are unable exactly to define and determine. To increase the human energy is, therefore, equivalent to increasing this product, and there are, as will readily be seen, only three ways possible to attain this result, which are illustrated in the above diagram. The first way, shown in the top figure, is to increase the mass (as indicated by the dotted circle), leaving the two opposing forces the same. The second way is to reduce the retarding force R to a smaller value r, leaving the mass and the impelling force the same, as diagrammatically shown in the middle figure. The third way, which is illustrated in the last figure, is to increase the impelling force f to a higher value F, while the mass and the retarding force R remain unaltered. Evidently fixed limits exist as regards increase of mass and reduction of retarding force, but the impelling force can be increased indefinitely. Each of these three possible solutions presents a different aspect of the main problem of increasing human energy, which is thus divided into three distinct problems, to be successively considered.

The First Problem: How to Increase the Human Mass—The Burning of Atmospheric Nitrogen.

Viewed generally, there are obviously two ways of increasing the mass of mankind: first, by aiding and maintaining those forces and conditions which tend to increase it; and, second, by opposing and reducing those which tend to diminish it. The mass will be increased by careful attention to health, by substantial food, by moderation, by regularity of habits, by the promotion of marriage, by conscientious attention to the children, and, generally stated, by the observance of all the many precepts and laws of religion and hygiene. But in adding new mass to the old, three cases again present themselves. Either the mass added is of the same velocity as the old, or it is of a smaller or of a higher velocity. To gain an idea of the relative importance of these cases, imagine a train composed of, say, one hundred locomotives running on a track, and suppose that, to increase the energy of the moving mass, four more locomotives are added to the train. If these four move at the same velocity at which the train is going, the total energy will be increased four per cent.; if they are moving at only one half of that velocity, the increase will amount to only one per cent.; if they are moving at twice that velocity, the increase of energy will be sixteen per cent. This simple illustration shows that it is of the greatest importance to add mass of a higher velocity. Stated more to the point, if, for example, the children be of the same degree of enlightenment as the parents,—that is, mass of the "same velocity,"—the energy will simply increase proportionately to the number added. If they are less intelligent or advanced, or mass of "smaller velocity," there will be a very slight gain in the energy; but if they are further advanced, or mass of "higher velocity," then the new generation will add very considerably to the sum total of human energy. Any addition of mass of "smaller velocity," beyond that indispensable amount required by the law expressed in the proverb, "Mens sana in corpore sano," should be strenuously opposed. For instance, the mere development of muscle, as aimed at in some of our colleges, I consider equivalent to adding mass of "smaller velocity," and I would not commend it, although my views were different when I was a student myself. Moderate exercise, insuring the right balance between mind and body, and the highest efficiency of performance, is, of course, a prime requirement. The above example shows that the most important result to be attained is the education, or the increase of the "velocity," of the mass newly added.

Conversely, it scarcely need be stated that everything that is against the teachings of religion and the laws of hygiene is tending to decrease the mass. Whisky, wine, tea, coffee, tobacco, and other such stimulants are responsible for the shortening of the lives of many, and ought to be used with moderation. But I do not think that rigorous measures of suppression of habits followed through many generations are commendable. It is wiser to preach moderation than abstinence. We have become accustomed to these stimulants, and if such reforms are to be effected, they must be slow and gradual. Those who are devoting their energies to such ends could make themselves far more useful by turning their efforts in other directions, as, for instance, toward providing pure water.

For every person who perishes from the effects of a stimulant, at least a thousand die from the consequences of drinking impure water. This precious fluid, which daily infuses new life into us, is likewise the chief vehicle through which disease and death enter our bodies. The germs of destruction it conveys are enemies all the more terrible as they perform their fatal work unperceived. They seal our doom while we live and enjoy. The majority of people are so ignorant or careless in drinking water, and the consequences of this are so disastrous, that a philanthropist can scarcely use his efforts better than by endeavoring to enlighten those who are thus injuring themselves. By systematic purification and sterilization of the drinking-water the human mass would be very considerably increased. It should be made a rigid rule—which might be enforced by law—to boil or to sterilize otherwise the drinking-water in every household and public place. The mere filtering does not afford sufficient security against infection. All ice for internal uses should be artificially prepared from water thoroughly sterilized. The importance of eliminating germs of disease from the city water is generally recognized, but little is being done to improve the existing conditions, as no satisfactory method of sterilizing great quantities of water has as yet been brought forward. By improved electrical appliances we are now enabled to produce ozone cheaply and in large amounts, and this ideal disinfectant seems to offer a happy solution of the important question.

Gambling, business rush, and excitement, particularly on the exchanges, are causes of much mass-reduction, all the more so because the individuals concerned represent units of higher value. Incapacity of observing the first symptoms of an illness, and careless neglect of the same, are important factors of mortality. In noting carefully every new sign of approaching danger, and making conscientiously every possible effort to avert it, we are not only following wise laws of hygiene in the interest of our well-being and the success of our labors, but we are also complying with a higher moral duty. Every one should consider his body as a priceless gift from one whom he loves above all, as a marvelous work of art, of undescribable beauty and mastery beyond human conception, and so delicate and frail that a word, a breath, a look, nay, a thought, may injure it. Uncleanliness, which breeds disease and death, is not only a self-destructive but a highly immoral habit. In keeping our bodies free from infection, healthful, and pure, we are expressing our reverence for the high principle with which they are endowed. He who follows the precepts of hygiene in this spirit is proving himself, so far, truly religious. Laxity of morals is a terrible evil, which poisons both mind and body, and which is responsible for a great reduction of the human mass in some countries. Many of the present customs and tendencies are productive of similar hurtful results. For example, the society life, modern education and pursuits of women, tending to draw them away from their household duties and make men out of them, must needs detract from the elevating ideal they represent, diminish the artistic creative power, and cause sterility and a general weakening of the race. A thousand other evils might be mentioned, but all put together, in their bearing upon the problem under discussion, they would not equal a single one, the want of food, brought on by poverty, destitution, and famine. Millions of individuals die yearly for want of food, thus keeping down the mass. Even in our enlightened communities, and notwithstanding the many charitable efforts, this is still, in all probability, the chief evil. I do not mean here absolute want of food, but want of healthful nutriment.

How to provide good and plentiful food is, therefore, a most important question of the day. On general principles the raising of cattle as a means of providing food is objectionable, because, in the sense interpreted above, it must undoubtedly tend to the addition of mass of a "smaller velocity." It is certainly preferable to raise vegetables, and I think, therefore, that vegetarianism is a commendable departure from the established barbarous habit. That we can subsist on plant food and perform our work even to advantage is not a theory, but a well-demonstrated fact. Many races living almost exclusively on vegetables are of superior physique and strength. There is no doubt that some plant food, such as oatmeal, is more economical than meat, and superior to it in regard to both mechanical and mental performance. Such food, moreover, taxes our digestive organs decidedly less, and, in making us more contented and sociable, produces an amount of good difficult to estimate. In view of these facts every effort should be made to stop the wanton and cruel slaughter of animals, which must be destructive to our morals. To free ourselves from animal instincts and appetites, which keep us down, we should begin at the very root from which they spring: we should effect a radical reform in the character of the food.

There seems to be no *philosophical* necessity for food. We can conceive of organized beings living without nourishment, and deriving all the energy they need for the performance of their life-functions from the ambient medium. In a crystal we have the clear evidence of the existence of a formative life-principle, and though we cannot understand the life of a crystal, it is none the less a living being. There may be, besides crystals, other such individualized, material systems of beings, perhaps of gaseous constitution, or composed of substance still more tenuous. In view of this possibility,—nay, probability,—we cannot apodictically deny the existence of organized beings on a planet merely because the conditions on the same are unsuitable for the existence of life as we conceive it. We cannot even, with positive assurance, assert that some of them might not be present here, in this our world, in the very midst of us, for their constitution and life-manifestation may be such that we are unable to perceive them.

The production of artificial food as a means for causing an increase of the human mass naturally suggests itself, but a direct attempt of this kind to provide nourishment does not appear to me rational, at least not for the present. Whether we could thrive on such food is very doubtful. We are the result of ages of continuous adaptation, and we cannot radically change without unforeseen and, in all probability, disastrous consequences. So uncertain an experiment should not be tried. By far the best way, it seems to me, to meet the ravages of the evil, would be to find ways of increasing the productivity of the soil. With this object the preservation of forests is of an importance which cannot be overestimated, and in this connection, also, the utilization of waterpower for purposes of electrical transmission, dispensing in many ways with the necessity of burning wood, and tending thereby to forest preservation, is to be strongly advocated. But there are limits in the improvement to be effected in this and similar ways.

To increase materially the productivity of the soil, it must be more effectively fertilized by artificial means. The question of food-production resolves itself, then, into the question how best to fertilize the soil. What it is that made the soil is still a mystery. To explain its origin is probably equivalent to explaining the origin of life itself. The rocks, disintegrated by moisture and heat and wind and weather, were in themselves not capable of maintaining life. Some unexplained condition arose, and some new principle came into effect, and the first layer capable of sustaining low organisms, like mosses, was formed. These, by their life and death, added more of the life-sustaining quality to the soil, and higher organisms could then subsist, and so on and on, until at last highly developed plant and animal life could flourish. But though the theories are, even now, not in agreement as to how fertilization is effected, it is a fact, only too well ascertained, that the soil cannot indefinitely sustain life, and some way must be found to supply it with the substances which have been abstracted from it by the plants. The chief and most valuable among these substances are compounds of nitrogen, and the cheap production of these is, therefore, the key for the solution of the all-important food problem.

Our atmosphere contains an inexhaustible amount of nitrogen, and could we but oxidize it and produce these compounds, an incalculable benefit for mankind would follow.

Long ago this idea took a powerful hold on the imagination of scientific men, but an efficient means for accomplishing this result could not be devised. The problem was rendered extremely difficult by the extraordinary inertness of the nitrogen, which refuses to combine even with oxygen. But here electricity comes to our aid: the dormant affinities of the element are awakened by an electric current of the proper quality. As a lump of coal which has been in contact with oxygen for centuries without burning will combine with it when once ignited, so nitrogen, excited by electricity, will burn. I did not succeed, however, in producing electrical discharges exciting very effectively the atmospheric nitrogen until a comparatively recent date, although I showed, in May, 1891, in a scientific lecture, a novel form of discharge or electrical flame named "St. Elmo's hotfire," which, besides being capable of generating ozone in abundance, also possessed, as I pointed out on that occasion, distinctly the quality of exciting chemical affinities. This discharge or flame was then only three or four inches long, its chemical action was likewise very feeble, and consequently the process of oxidation of the nitrogen was wasteful. How to intensify this action was the question. Evidently electric currents of a peculiar kind had to be produced in order to render the process of nitrogen combustion more efficient.

The first advance was made in ascertaining that the chemical activity of the discharge was very considerably increased by using currents of extremely high frequency or rate of vibration. This was an important improvement, but practical considerations soon set a definite limit to the progress in this direction. Next, the effects of the electrical pressure of the current impulses, of their wave-form and other characteristic features, were investigated. Then the influence of the atmospheric pressure and temperature and of the presence of water and other bodies was studied, and thus the best conditions for causing the most intense chemical action of the discharge and securing the highest efficiency of the process were gradually ascertained. Naturally, the improvements were not quick in coming; still, little by little, I advanced. The flame grew larger and larger, and its oxidizing action more and more intense. From an insignificant brush-discharge a few inches long it developed into a marvelous electrical phenomenon, a roaring blaze, devouring the nitrogen of the atmosphere and measuring sixty or seventy feet across. Thus slowly, almost imperceptibly, possibility became accomplishment. All is not yet done, by any means, but to what a degree my efforts have been rewarded an idea may be gained from an inspection of Fig. 1, which, with its title, is self-explanatory. The flame-like discharge visible is produced by the intense electrical oscillations which pass through the coil shown, and violently agitate the electrified molecules of the air. By this means a strong affinity is created between the two normally indifferent constituents of the atmosphere, and they combine readily, even if no further provision is made for intensifying the chemical action of the discharge. In the manufacture of nitrogen compounds by this method, of course, every possible means bearing upon the intensity of this action and the efficiency of the process will be taken advantage of, and, besides, special arrangements will be provided for the fixation of the compounds formed, as they are generally unstable, the nitrogen becoming again inert after a little lapse of time. Steam is a simple and effective means for fixing permanently the compounds. The result illustrated makes it practicable to oxidize the atmospheric nitrogen in unlimited quantities, merely by the use of cheap mechanical power and simple electrical apparatus. In this manner many compounds of nitrogen may be manufactured all over the world, at a small cost, and in any desired amount, and by means of these compounds the soil can be fertilized and its productiveness indefinitely increased. An abundance of cheap and healthful food, not artificial, but such as we are accustomed to, may thus be obtained: This new and inexhaustible source of food-supply will be of incalculable benefit to mankind, for it will enormously contribute to the increase of the human mass, and thus add immensely to human energy. Soon, I hope, the world will see the beginning of an industry which, in time to come, will, I believe, be in importance next to that of iron.

The Second Problem: How to Reduce the Force Retarding the Human Mass—The Art of Telautomatics.

As before stated, the force which retards the onward movement of man is partly frictional and partly negative. To illustrate this distinction I may name, for example, ignorance, stupidity, and imbecility as some of the purely frictional forces, or resistances devoid of any directive tendency. On the other hand, visionariness, insanity, self-destructive tendency, religious fanaticism, and the like, are all forces of a negative character, acting in definite directions. To reduce or entirely to overcome these dissimilar retarding forces, radically different methods must be employed. One knows, for instance, what a fanatic may do, and one can take preventive measures, can enlighten, convince, and possibly direct him, turn his vice into virtue; but one does not know, and never can know, what a brute or an imbecile may do, and one must deal with him as with a mass, inert, without mind, let loose by the mad elements. A negative force always implies some quality, not infrequently a high one, though badly directed, which it is possible to turn to good advantage; but a directionless, frictional force involves unavoidable loss. Evidently, then, the first and general answer to the above question is: turn all negative force in the right direction and reduce all frictional force.

There can be no doubt that, of all the frictional resistances, the one that most retards human movement is ignorance. Not without reason said that man of wisdom, Buddha: "Ignorance is the greatest evil in the world." The friction which results from ignorance, and which is greatly increased owing to the numerous languages and nationalities, can be reduced only by the spread of knowledge and the unification of the heterogeneous elements of humanity. No effort could be better spent.

But however ignorance may have retarded the onward movement of man in times past, it is certain that, nowadays, negative forces have become of greater importance. Among these there is one of far greater moment than any other. It is called organized warfare. When we consider the millions of individuals, often the ablest in mind and body, the flower of humanity, who are compelled to a life of inactivity and unproductiveness, the immense sums of money daily required for the maintenance of armies and war apparatus, representing ever so much of human energy, all the effort uselessly spent in the production of arms and implements of destruction, the loss of life and the fostering of a barbarous spirit, we are appalled at the inestimable loss to mankind which the existence of these deplorable conditions must involve. What can we do to combat best this great evil?

Law and order absolutely require the maintenance of organized force. No community can exist and prosper without rigid discipline. Every country must be able to defend itself, should the necessity arise. The conditions of to-day are not the result of yesterday, and a radical change cannot be effected to-morrow. If the nations would at once disarm, it is more than likely that a state of things worse than war itself would follow. Universal peace is a beautiful dream, but not at once realizable. We have seen recently that even the noble effort of the man invested with the greatest worldly power has been virtually without effect. And no wonder, for the establishment of universal peace is, for the time being, a physical impossibility. War is a negative force, and cannot be turned in a positive direction without passing through the intermediate phases. It is the problem of making a wheel, rotating one way, turn in the opposite direction without slowing it down, stopping it, and speeding it up again the other way.

It has been argued that the perfection of guns of great destructive power will stop warfare. So I myself thought for a long time, but now I believe this to be a profound mistake. Such developments will greatly modify, but not arrest it. On the contrary, I think that every new arm that is invented, every new departure that is made in this direction, merely invites new talent and skill, engages new effort, offers a new incentive, and so only gives a fresh impetus to further development. Think of the discovery of gunpowder. Can we conceive of any more radical departure than was effected by this innovation? Let us imagine ourselves living in that period: would we not have thought then that warfare was at an end, when the armor of the knight became an object of ridicule, when bodily strength and skill, meaning so much before, became of comparatively little value? Yet gunpowder did not stop warfare; quite the opposite—it acted as a most powerful incentive. Nor do I believe that warfare can ever be arrested by any scientific or ideal development, so long as similar conditions to those now prevailing exist, because war has itself become a science, and because war involves some of the most sacred sentiments of which man is capable. In fact, it is doubtful whether men who would not be ready to fight for a high principle would be good for anything at all. It is not the mind which makes man, nor is it the body; it is mind and body. Our virtues and our failings are inseparable, like force and matter. When they separate, man is no more.

Another argument, which carries considerable force, is frequently made, namely, that war must soon become impossible because the means of defense are outstripping the means of attack. This is only in accordance with a fundamental law which may be expressed by the statement that it is easier to destroy than to build. This law defines human capacities and human conditions. Were these such that it would be easier to build than to destroy, man would go on unresisted, creating and accumulating without limit. Such conditions are not of this earth. A being which could do this would not be a man; it might be a god. Defense will always have the advantage over attack, but this alone, it seems to me, can never stop war. By the use of new principles of defense we can render harbors impregnable against attack, but we cannot by such means prevent two war-ships meeting in battle on the high sea. And then, if we follow this idea to its ultimate development, we are led to the conclusion that it would be better for mankind if attack and defense were just oppositely related; for if every country, even the smallest, could surround itself with a wall absolutely impenetrable, and could defy the rest of the world, a state of things would surely be brought on which would be extremely unfavorable to human progress. It is by abolishing all the barriers which separate nations and countries that civilization is best furthered.

Again, it is contended by some that the advent of the flying-machine must bring on universal peace. This, too, I believe to be an entirely erroneous view. The flying-machine is certainly coming, and very soon, but the conditions will remain the same as before. In fact, I see no reason why a ruling power, like Great Britain, might not govern the air as well as the sea. Without wishing to put myself on record as a prophet, I do not hesitate to say that the next years will see the establishment of an "air-power," and its center may not be far from New York. But, for all that, men will fight on merrily.

The ideal development of the war principle would ultimately lead to the transformation of the whole energy of war into purely potential, explosive energy, like that of an electrical condenser. In this form the war-energy could be maintained without effort; it would need to be much smaller in amount, while incomparably more effective.

As regards the security of a country against foreign invasion, it is interesting to note that it depends only on the relative, and not on the absolute, number of the individuals or magnitude of the forces, and that, if every country should reduce the war-force in the same ratio, the security would remain unaltered. An international agreement with the object of reducing to a minimum the war-force which, in view of the present still imperfect education of the masses, is absolutely indispensable, would, therefore, seem to be the first rational step to take toward diminishing the force retarding human movement.

Fortunately, the existing conditions cannot continue indefinitely, for a new element is beginning to assert itself. A change for the better is imminent, and I shall now endeavor to show what, according to my ideas, will be the first advance toward the establishment of peaceful relations between nations, and by what means it will eventually be accomplished.

Let us go back to the early beginning, when the law of the stronger was the only law. The light of reason was not yet kindled, and the weak was entirely at the mercy of the strong. The weak individual then began to learn how to defend himself. He made use of a club, stone, spear, sling, or bow and arrow, and in the course of time, instead of physical strength, intelligence became the chief deciding factor in the battle. The wild character was gradually softened by the awakening of noble sentiments, and so, imperceptibly, after ages of continued progress, we have come from the brutal fight of the unreasoning animal to what we call the "civilized warfare" of to-day, in which the combatants shake hands, talk in a friendly way, and smoke cigars in the entr'actes, ready to engage again in deadly conflict at a signal. Let pessimists say what they like, here is an absolute evidence of great and gratifying advance.

But now, what is the next phase in this evolution? Not peace as yet, by any means. The next change which should naturally follow from modern developments should be the continuous diminution of the number of individuals engaged in battle. The apparatus will be one of specifically great power, but only a few individuals will be required to operate it. This evolution will bring more and more into prominence a machine or mechanism with the fewest individuals as an element of warfare, and the absolutely unavoidable consequence of this will be the abandonment of large, clumsy, slowly moving, and unmanageable units. Greatest possible speed and maximum rate of energy-delivery by the war apparatus will be the main object. The loss of life will become smaller and smaller, and finally, the number of the individuals continuously diminishing, merely machines will meet in a contest without bloodshed, the nations being simply interested, ambitious spectators. When this happy condition is realized, peace will be assured. But, no matter to what degree of perfection rapid-fire guns, high-power cannon, explosive projectiles, torpedo-boats, or other implements of war may be brought, no matter how destructive they may be made, that condition can never be reached through any such development. All such implements require men for their operation; men are indispensable parts of the machinery. Their object is to kill and to destroy. Their power resides in their capacity for doing evil. So long as men meet in battle, there will be bloodshed. Bloodshed will ever keep up barbarous passion. To break this fierce spirit, a radical departure must be made, an entirely new principle must be introduced, something that never existed before in warfare—a principle which will forcibly, unavoidably, turn the battle into a mere spectacle, a play, a contest without loss of blood. To bring on this result men must be dispensed with: machine must fight machine. But how accomplish that which seems impossible? The answer is simple enough: produce a machine capable of acting as though it were part of a human being—no mere mechanical contrivance, comprising levers, screws, wheels, clutches, and nothing more, but a machine embodying a higher principle, which will enable it to perform its duties as though it had intelligence, experience, reason, judgment, a mind! This conclusion is the result of my thoughts and observations which have extended through virtually my whole life, and I shall now briefly describe how I came to accomplish that which at first seemed an unrealizable dream.

A long time ago, when I was a boy, I was afflicted with a singular trouble, which seems to have been due to an extraordinary excitability of the retina. It was the appearance of images which, by their persistence, marred the vision of real objects and interfered with thought. When a word was said to me, the image of the object which it designated would appear vividly before my eyes, and many times it was impossible for me to tell whether the object I saw was real or not. This caused me great discomfort and anxiety, and I tried hard to free myself of the spell. But for a long time I tried in vain, and it was not, as I still clearly recollect, until I was about twelve years old that I succeeded for the first time, by an effort of the will, in banishing an image which presented itself. My happiness will never be as complete as it was then, but, unfortunately (as I thought at that time), the old trouble returned, and with it my anxiety.

Here it was that the observations to which I refer began. I noted, namely, that whenever the image of an object appeared before my eyes I had seen something which reminded me of it. In the first instances I thought this to be purely accidental, but soon I convinced myself that it was not so. A visual impression, consciously or unconsciously received, invariably preceded the appearance of the image. Gradually the desire arose in me to find out, every time, what caused the images to appear, and the satisfaction of this desire soon became a necessity. The next observation I made was that, just as these images followed as a result of something I had seen, so also the thoughts which I conceived were suggested in like manner. Again, I experienced the same desire to locate the image which caused the thought, and this search for the original visual impression soon grew to be a second nature. My mind became automatic, as it were, and in the course of years of continued, almost unconscious performance, I acquired the ability of locating every time and, as a rule, instantly the visual impression which started the thought. Nor is this all. It was not long before I was aware that also all my movements were prompted in the same way, and so, searching, observing, and verifying continuously, year after year, I have, by every thought and every act of mine, demonstrated, and do so daily, to my absolute satisfaction, that I am an automaton endowed with power of movement, which merely responds to external stimuli beating upon my sense organs, and thinks and acts and moves accordingly. I remember only one or two cases in all my life in which I was unable to locate the first impression which prompted a movement or a thought, or even a dream.

With these experiences it was only natural that, long ago, I conceived the idea of constructing an automaton which would mechanically represent me, and which would respond, as I do myself, but, of course, in a much more primitive manner, to external influences. Such an automaton evidently had to have motive power, organs for locomotion, directive organs, and one or more sensitive organs so adapted as to be excited by external stimuli. This machine would, I reasoned, perform its movements in the manner of a living being, for it would have all the chief mechanical characteristics or elements of the same. There was still the capacity for growth, propagation, and, above all, the mind which would be wanting to make the model complete. But growth was not necessary in this case, since a machine could be manufactured full-grown, so to speak. As to the capacity for propagation, it could likewise be left out of consideration, for in the mechanical model it merely signified a process of manufacture. Whether the automaton be of flesh and bone, or of wood and steel, it mattered little, provided it could perform all the duties required of it like an intelligent being. To do so, it had to have an element corresponding to the mind, which would effect the control of all its movements and operations, and cause it to act, in any unforeseen case that might present itself, with knowledge, reason, judgment, and experience. But this element I could easily embody in it by conveying to it my own intelligence, my own understanding. So this invention was evolved, and so a new art came into existence, for which the name "telautomatics" has been suggested, which means the art of controlling the movements and operations of distant automatons.

This principle evidently was applicable to any kind of machine that moves on land or in the water or in the air. In applying it practically for the first time, I selected a boat (see Fig. 2). A storage battery placed within it furnished the motive power. The propeller, driven by a motor, represented the locomotive organs. The rudder, controlled by another motor likewise driven by the battery, took the place of the directive organs. As to the sensitive organ, obviously the first thought was to utilize a device responsive to rays of light, like a selenium cell, to represent the human eye. But upon closer inquiry I found that, owing to experimental and other difficulties, no thoroughly satisfactory control of the automaton could be effected by light, radiant heat, Hertzian radiations, or by rays in general, that is, disturbances which pass in straight lines through space. One of the reasons was that any obstacle coming between the operator and the distant automaton would place it beyond his control. Another reason was that the sensitive device representing the eye would have to be in a definite position with respect to the distant controlling apparatus, and this necessity would impose great limitations in the control. Still another and very important reason was that, in using rays, it would be difficult, if not impossible, to give to the automaton individual features or characteristics distinguishing it from other machines of this kind. Evidently the automaton should respond only to an individual call, as a person responds to a name. Such considerations led me to conclude that the sensitive device of the machine should correspond to the ear rather than to the eye of a human being, for in this case its actions could be controlled irrespective of intervening obstacles, regardless of its position relative to the distant controlling apparatus, and, last, but not least, it would remain deaf and unresponsive, like a faithful servant, to all calls but that of its master. These requirements made it imperative to use, in the control of the automaton, instead of light- or other rays, waves or disturbances which propagate in all directions through space, like sound, or which follow a path of least resistance, however curved. I attained the result aimed at by means of an electric circuit placed within the boat, and adjusted, or "tuned," exactly to electrical vibrations of the proper kind transmitted to it from a distant "electrical oscillator." This circuit, in responding, however feebly, to the transmitted vibrations, affected magnets and other contrivances, through the medium of which were controlled the movements of the propeller and rudder, and also the operations of numerous other appliances.

By the simple means described the knowledge, experience, judgment—the mind, so to speak—of the distant operator were embodied in that machine, which was thus enabled to move and to perform all its operations with reason and intelligence. It behaved just like a blindfolded person obeying directions received through the ear.

The automatons so far constructed had "borrowed minds," so to speak, as each merely formed part of the distant operator who conveyed to it his intelligent orders; but this art is only in the beginning. I purpose to show that, however impossible it may now seem, an automaton may be contrived which will have its "own mind," and by this I mean that it will be able, independent of any operator, left entirely to itself, to perform, in response to external influences affecting its sensitive organs, a great variety of acts and operations as if it had intelligence. It will be able to follow a course laid out or to obey orders given far in advance; it will be capable of distinguishing between what it ought and what it ought not to do, and of making experiences or, otherwise stated, of recording impressions which will definitely affect its subsequent actions. In fact, I have already conceived such a plan.

Although I evolved this invention many years ago and explained it to my visitors very frequently in my laboratory demonstrations, it was not until much later, long after I had perfected it, that it became known, when, naturally enough, it gave rise to much discussion and to sensational reports. But the true significance of this new art was not grasped by the majority, nor was the great force of the underlying principle recognized. As nearly as I could judge from the numerous comments which then appeared, the results I had obtained were considered as entirely impossible. Even the few who were disposed to admit the practicability of the invention saw in it merely an automobile torpedo, which was to be used for the purpose of blowing up battle-ships, with doubtful success. The general impression was that I contemplated simply the steering of such a vessel by means of Hertzian or other rays. There are torpedoes steered electrically by wires, and there are means of communicating without wires, and the above was, of course, an obvious inference. Had I accomplished nothing more than this, I should have made a small advance indeed. But the art I have evolved does not contemplate merely the change of direction of a moving vessel; it affords a means of absolutely controlling, in every respect, all the innumerable translatory movements, as well as the operations of all the internal organs, no matter how many, of an individualized automaton. Criticisms to the effect that the control of the automaton could be interfered with were made by people who do not even dream of the wonderful results which can be accomplished by the use of electrical vibrations. The world moves slowly, and new truths are difficult to see. Certainly, by the use of this principle, an arm for attack as well as defense may be provided, of a destructiveness all the greater as the principle is applicable to submarine and aërial vessels. There is virtually no restriction as to the amount of explosive it can carry, or as to the distance at which it can strike, and failure is almost impossible. But the force of this new principle does not wholly reside in its destructiveness. Its advent introduces into warfare an element which never existed before—a fighting-machine without men as a means of attack and defense. The continuous development in this direction must ultimately make war a mere contest of machines without men and without loss of life—a condition which would have been impossible without this new departure, and which, in my opinion, must be reached as preliminary to permanent peace. The future will either bear out or disprove these views. My ideas on this subject have been put forth with deep conviction, but in a humble spirit.

The establishment of permanent peaceful relations between nations would most effectively reduce the force retarding the human mass, and would be the best solution of this great human problem. But will the dream of universal peace ever be realized? Let us hope that it will. When all darkness shall be dissipated by the light of science, when all nations shall be merged into one, and patriotism shall be identical with religion, when there shall be one language, one country, one end, then the dream will have become reality.

The Third Problem: How to Increase the Force Accelerating the Human Mass—The Harnessing of the Sun's Energy.

Of the three possible solutions of the main problem of increasing human energy, this is by far the most important to consider, not only because of its intrinsic significance, but also because of its intimate bearing on all the many elements and conditions which determine the movement of humanity. In order to proceed systematically, it would be necessary for me to dwell on all those considerations which have guided me from the outset in my efforts to arrive at a solution, and which have led me, step by step, to the results I shall now describe. As a preliminary study of the problem an analytical investigation, such as I have made, of the chief forces which determine the onward movement, would be of advantage, particularly in conveying an idea of that hypothetical "velocity" which, as explained in the beginning, is a measure of human energy; but to deal with this specifically here, as I would desire, would lead me far beyond the scope of the present subject. Suffice it to state that the resultant of all these forces is always in the direction of reason, which, therefore, determines, at any time, the direction of human movement. This is to say that every effort which is scientifically applied, rational, useful, or practical, must be in the direction in which the mass is moving. The practical, rational man, the observer, the man of business, he who reasons, calculates, or determines in advance, carefully applies his effort so that when coming into effect it will be in the direction of the movement, making it thus most efficient, and in this knowledge and ability lies the secret of his success. Every new fact discovered, every new experience or new element added to our knowledge and entering into the domain of reason, affects the same and, therefore, changes the direction of the movement, which, however, must always take place along the resultant of all those efforts which, at that time, we designate as reasonable, that is, self-preserving, useful, profitable, or practical. These efforts concern our daily life, our necessities and comforts, our work and business, and it is these which drive man onward.

But looking at all this busy world about us, on all this complex mass as it daily throbs and moves, what is it but an immense clockwork driven by a spring? In the morning, when we rise, we cannot fail to note that all the objects about us are manufactured by machinery: the water we use is lifted by steam-power; the trains bring our breakfast from distant localities; the elevators in our dwelling and in our office building, the cars that carry us there, are all driven by power; in all our daily errands, and in our very life-pursuit, we depend upon it; all the objects we see tell us of it; and when we return to our machine-made dwelling at night, lest we should forget it, all the material comforts of our home, our cheering stove and lamp, remind us how much we depend on power. And when there is an accidental stoppage of the machinery, when the city is snow-bound, or the life-sustaining movement otherwise temporarily arrested, we are affrighted to realize how impossible it would be for us to live the life we live without motive power. Motive power means work. To increase the force accelerating human movement means, therefore, to perform more work.

So we find that the three possible solutions of the great problem of increasing human energy are answered by the three words: *food, peace, work.* Many a year I have thought and pondered, lost myself in speculations and theories, considering man as a mass moved by a force, viewing his inexplicable movement in the light of a mechanical one, and applying the simple principles of mechanics to the analysis of the same until I arrived at these solutions, only to realize that they were taught to me in my early childhood. These three words sound the key-notes of the Christian religion. Their scientific meaning and purpose are now clear to me: food to increase the mass, peace to diminish the retarding force, and work to increase the force accelerating human movement. These are the only three solutions which are possible of that great problem, and all of them have one object, one end, namely, to increase human energy. When we recognize this, we cannot help wondering how profoundly wise and scientific and how immensely practical the Christian religion is, and in what a marked contrast it stands in this respect to other religions. It is unmistakably the result of practical experiment and scientific observation which have extended through ages, while other religions seem to be the outcome of merely abstract reasoning. Work, untiring effort, useful and accumulative, with periods of rest and recuperation aiming at higher efficiency, is its chief and ever-recurring command. Thus we are inspired both by Christianity and Science to do our utmost toward increasing the performance of mankind. This most important of human problems I shall now specifically consider.

The Source of Human Energy—
The Three Ways of Drawing
Energy from the Sun.

First let us ask: Whence comes all the motive power? What is the spring that drives all? We see the ocean rise and fall, the rivers flow, the wind, rain, hail, and snow beat on our windows, the trains and steamers come and go; we hear the rattling noise of carriages, the voices from the street; we feel, smell, and taste; and we think of all this. And all this movement, from the surging of the mighty ocean to that subtle movement concerned in our thought, has but one common cause. All this energy emanates from one single center, one single source— the sun. The sun is the spring that drives all. The sun maintains all human life and supplies all human energy. Another answer we have now found to the above great question: To increase the force accelerating human movement means to turn to the uses of man more of the sun's energy. We honor and revere those great men of bygone times whose names are linked with immortal achievements, who have proved themselves benefactors of humanity—the religious reformer with his wise maxims of life, the philosopher with his deep truths, the mathematician with his formulæ, the physicist with his laws, the discoverer with his principles and secrets wrested from nature, the artist with his forms of the beautiful; but who honors him, the greatest of all,—who can tell the name of him,— who first turned to use the sun's energy to save the effort of a weak fellow-creature? That was man's first act of scientific philanthropy, and its consequences have been incalculable.

From the very beginning three ways of drawing energy from the sun were open to man. The savage, when he warmed his frozen limbs at a fire kindled in some way, availed himself of the energy of the sun stored in the burning material. When he carried a bundle of branches to his cave and burned them there, he made use of the sun's stored energy transported from one to another locality. When he set sail to his canoe, he utilized the energy of the sun supplied to the atmosphere or ambient medium. There can be no doubt that the first is the oldest way. A fire, found accidentally, taught the savage to appreciate its beneficial heat. He then very likely conceived the idea of carrying the glowing embers to his abode. Finally he learned to use the force of a swift current of water or air. It is characteristic of modern development that progress has been effected in the same order. The utilization of the energy stored in wood or coal; or, generally speaking, fuel, led to the steam-engine. Next a great stride in advance was made in energy-transportation by the use of electricity, which permitted the transfer of energy from one locality to another without transporting the material. But as to the utilization of the energy of the ambient medium, no radical step forward has as yet been made known.

The ultimate results of development in these three directions are: first, the burning of coal by a cold process in a battery; second, the efficient utilization of the energy of the ambient medium; and, third, the transmission without wires of electrical energy to any distance. In whatever way these results may be arrived at, their practical application will necessarily involve an extensive use of iron, and this invaluable metal will undoubtedly be an essential element in the further development along these three lines. If we succeed in burning coal by a cold process and thus obtaining electrical energy in an efficient and inexpensive manner, we shall require in many practical uses of this energy electric motors—that is, iron. If we are successful in deriving energy from the ambient medium, we shall need, both in the obtainment and utilization of the energy, machinery—again, iron. If we realize the transmission of electrical energy without wires on an industrial scale, we shall be compelled to use extensively electric generators—once more, iron. Whatever we may do, iron will probably be the chief means of accomplishment in the near future, possibly more so than in the past. How long its reign will last is difficult to tell, for even now aluminium is looming up as a threatening competitor. But for the time being, next to providing new resources of energy, it is of the greatest importance to make improvements in the manufacture and utilization of iron. Great advances are possible in these latter directions, which, if brought about, would enormously increase the useful performance of mankind.

Great Possibilities offered by Iron for Increasing Human Performance—Enormous Waste in Iron Manufacture.

Iron is by far the most important factor in modern progress. It contributes more than any other industrial product to the force accelerating human movement. So general is the use of this metal, and so intimately is it connected with all that concerns our life, that it has become as indispensable to us as the very air we breathe. Its name is synonymous with usefulness. But, however great the influence of iron may be on the present human development, it does not add to the force urging man onward nearly as much as it might. First of all, its manufacture as now carried on is connected with an appalling waste of fuel—that is, waste of energy. Then, again, only a part of all the iron produced is applied for useful purposes. A good part of it goes to create frictional resistances, while still another large part is the means of developing negative forces greatly retarding human movement. Thus the negative force of war is almost wholly represented in iron. It is impossible to estimate with any degree of accuracy the magnitude of this greatest of all retarding forces, but it is certainly very considerable. If the present positive impelling force due to all useful applications of iron be represented by ten, for instance, I should not think it exaggeration to estimate the negative force of war, with due consideration of all its retarding influences and results, at, say, six. On the basis of this estimate the effective impelling force of iron in the positive direction would be measured by the difference of these two numbers, which is four. But if, through the establishment of universal peace, the manufacture of war machinery should cease, and all struggle for supremacy between nations should be turned into healthful, ever active and productive commercial competition, then the positive impelling force due to iron would be measured by the sum of those two numbers, which is sixteen—that is, this force would have four times its present value. This example is, of course, merely intended to give an idea of the immense increase in the useful performance of mankind which would result from a radical reform of the iron industries supplying the implements of warfare.

A similar inestimable advantage in the saving of energy available to man would be secured by obviating the great waste of coal which is inseparably connected with the present methods of manufacturing iron. In some countries, as in Great Britain, the hurtful effects of this squandering of fuel are beginning to be felt. The price of coal is constantly rising, and the poor are made to suffer more and more. Though we are still far from the dreaded "exhaustion of the coal-fields," philanthropy commands us to invent novel methods of manufacturing iron, which will not involve such barbarous waste of this valuable material from which we derive at present most of our energy. It is our duty to coming generations to leave this store of energy intact for them, or at least not to touch it until we shall have perfected processes for burning coal more efficiently. Those who are to come after us will need fuel more than we do. We should be able to manufacture the iron we require by using the sun's energy, without wasting any coal at all. As an effort to this end the idea of smelting iron ores by electric currents obtained from the energy of falling water has naturally suggested itself to many. I have myself spent much time in endeavoring to evolve such a practical process, which would enable iron to be manufactured at small cost. After a prolonged investigation of the subject, finding that it was unprofitable to use the currents generated directly for smelting the ore, I devised a method which is far more economical.

Economical Production of Iron by a New Process.

The industrial project, as I worked it out six years ago, contemplated the employment of the electric currents derived from the energy of a waterfall, not directly for smelting the ore, but for decomposing water, as a preliminary step. To lessen the cost of the plant, I proposed to generate the currents in exceptionally cheap and simple dynamos, which I designed for this sole purpose. The hydrogen liberated in the electrolytic decomposition was to be burned or recombined with oxygen, not with that from which it was separated, but with that of the atmosphere. Thus very nearly the total electrical energy used up in the decomposition of the water would be recovered in the form of heat resulting from the recombination of the hydrogen. This heat was to be applied to the smelting of the ore. The oxygen gained as a by-product in the decomposition of the water I intended to use for certain other industrial purposes, which would probably yield good financial returns, inasmuch as this is the cheapest way of obtaining this gas in large quantities. In any event, it could be employed to burn all kinds of refuse, cheap hydrocarbon, or coal of the most inferior quality which could not be burned in air or be otherwise utilized to advantage, and thus again a considerable amount of heat would be made available for the smelting of the ore. To increase the economy of the process I contemplated, furthermore, using an arrangement such that the hot metal and the products of combustion, coming out of the furnace, would give up their heat upon the cold ore going into the furnace, so that comparatively little of the heat-energy would be lost in the smelting. I calculated that probably forty thousand pounds of iron could be produced per horse-power per annum by this method. Liberal allowances were made for those losses which are unavoidable, the above quantity being about half of that theoretically obtainable. Relying on this estimate and on practical data with reference to a certain kind of sand ore existing in abundance in the region of the Great Lakes, including cost of transportation and labor, I found that in some localities iron could be manufactured in this manner cheaper than by any of the adopted methods. This result would be attained all the more surely if the oxygen obtained from the water, instead of being used for smelting the ore, as assumed, should be more profitably employed. Any new demand for this gas would secure a higher revenue from the plant, thus cheapening the iron. This project was advanced merely in the interest of industry. Some day, I hope, a beautiful industrial butterfly will come out of the dusty and shriveled chrysalis.

The production of iron from sand ores by a process of magnetic separation is highly commendable in principle, since it involves no waste of coal; but the usefulness of this method is largely reduced by the necessity of melting the iron afterward. As to the crushing of iron ore, I would consider it rational only if done by water-power, or by energy otherwise obtained without consumption of fuel. An electrolytic cold process, which would make it possible to extract iron cheaply, and also to mold it into the required forms without any fuel consumption, would, in my opinion, be a very great advance in iron manufacture. In common with some other metals, iron has so far resisted electrolytic treatment, but there can be no doubt that such a cold process will ultimately replace in metallurgy the present crude method of casting, and thus obviate the enormous waste of fuel necessitated by the repeated heating of metal in the foundries.

Up to a few decades ago the usefulness of iron was based almost wholly on its remarkable mechanical properties, but since the advent of the commercial dynamo and electric motor its value to mankind has been greatly increased by its unique magnetic qualities. As regards the latter, iron has been greatly improved of late. The signal progress began about thirteen years ago, when I discovered that in using soft Bessemer steel instead of wrought iron, as then customary, in an alternating motor, the performance of the machine was doubled. I brought this fact to the attention of Mr. Albert Schmid, to whose untiring efforts and ability is largely due the supremacy of American electrical machinery, and who was then superintendent of an industrial corporation engaged in this field. Following my suggestion, he constructed transformers of steel, and they showed the same marked improvement. The investigation was then systematically continued under Mr. Schmid's guidance, the impurities being gradually eliminated from the "steel" (which was only such in name, for in reality it was pure soft iron), and soon a product resulted which admitted of little further improvement.

The Coming Age of Aluminium—
Doom of the Copper Industry—
The Great Civilizing Potency of
The New Metal.

With the advances made in iron of late years we have arrived virtually at the limits of improvement. We cannot hope to increase very materially its tensile strength, elasticity, hardness, or malleability, nor can we expect to make it much better as regards its magnetic qualities. More recently a notable gain was secured by the mixture of a small percentage of nickel with the iron, but there is not much room for further advance in this direction. New discoveries may be expected, but they cannot greatly add to the valuable properties of the metal, though they may considerably reduce the cost of manufacture. The immediate future of iron is assured by its cheapness and its unrivaled mechanical and magnetic qualities. These are such that no other product can compete with it now. But there can be no doubt that, at a time not very distant, iron, in many of its now uncontested domains, will have to pass the scepter to another: the coming age will be the age of aluminium. It is only seventy years since this wonderful metal was discovered by Woehler, and the aluminium industry, scarcely forty years old, commands already the attention of the entire world. Such rapid growth has not been recorded in the history of civilization before. Not long ago aluminium was sold at the fanciful price of thirty or forty dollars per pound; to-day it can be had in any desired amount for as many cents. What is more, the time is not far off when this price, too, will be considered fanciful, for great improvements are possible in the methods of its manufacture. Most of the metal is now produced in the electric furnace by a process combining fusion and electrolysis, which offers a number of advantageous features, but involves naturally a great waste of the electrical energy of the current. My estimates show that the price of aluminium could be considerably reduced by adopting in its manufacture a method similar to that proposed by me for the production of iron. A pound of aluminium requires for fusion only about seventy per cent of the heat needed for melting a pound of iron, and inasmuch as its weight is only about one third of that of the latter, a volume of aluminium four times that of iron could be obtained from a given amount of heat-energy. But a cold electrolytic process of manufacture is the ideal solution, and on this I have placed my hope.

The absolutely unavoidable consequence of the advance of the aluminium industry will be the annihilation of the copper industry. They cannot exist and prosper together, and the latter is doomed beyond any hope of recovery. Even now it is cheaper to convey an electric current through aluminium wires than through copper wires: aluminium castings cost less, and in many domestic and other uses copper has no chance of successfully competing. A further material reduction of the price of aluminium cannot but be fatal to copper. But the progress of the former will not go on unchecked, for, as it ever happens in such cases, the larger industry will absorb the smaller one: the giant copper interests will control the pygmy aluminium interests, and the slow-pacing copper will reduce the lively gait of aluminium. This will only delay, not avoid, the impending catastrophe.

Aluminium, however, will not stop at downing copper. Before many years have passed it will be engaged in a fierce struggle with iron, and in the latter it will find an adversary not easy to conquer. The issue of the contest will largely depend on whether iron shall be indispensable in electric machinery. This the future alone can decide. The magnetism as exhibited in iron is an isolated phenomenon in nature. What it is that makes this metal behave so radically different from all other materials in this respect has not yet been ascertained, though many theories have been suggested. As regards magnetism, the molecules of the various bodies behave like hollow beams partly filled with a heavy fluid and balanced in the middle in the manner of a see-saw. Evidently some disturbing influence exists in nature which causes each molecule, like such a beam, to tilt either one or the other way. If the molecules are tilted one way, the body is magnetic; if they are tilted the other way, the body is non-magnetic; but both positions are stable, as they would be in the case of the hollow beam, owing to the rushing of the fluid to the lower end. Now, the wonderful thing is that the molecules of all known bodies went one way, while those of iron went the other way. This metal, it would seem, has an origin entirely different from that of the rest of the globe. It is highly improbable that we shall discover some other and cheaper material which will equal or surpass iron in magnetic qualities.

Unless we should make a radical departure in the character of the electric currents employed, iron will be indispensable. Yet the advantages it offers are only apparent. So long as we use feeble magnetic forces it is by far superior to any other material; but if we find ways of producing great magnetic forces, then better results will be obtainable without it. In fact, I have already produced electric transformers in which no iron is employed, and which are capable of performing ten times as much work per pound of weight as those with iron. This result is attained by using electric currents of a very high rate of vibration, produced in novel ways, instead of the ordinary currents now employed in the industries. I have also succeeded in operating electric motors without iron by such rapidly vibrating currents, but the results, so far, have been inferior to those obtained with ordinary motors constructed of iron, although theoretically the former should be capable of performing incomparably more work per unit of weight than the latter. But the seemingly insuperable difficulties which are now in the way may be overcome in the end, and then iron will be done away with, and all electric machinery will be manufactured of aluminium, in all probability, at prices ridiculously low. This would be a severe, if not a fatal, blow to iron. In many other branches of industry, as ship-building, or wherever lightness of structure is required, the progress of the new metal will be much quicker. For such uses it is eminently suitable, and is sure to supersede iron sooner or later. It is highly probable that in the course of time we shall be able to give it many of those qualities which make iron so valuable.

While it is impossible to tell when this industrial revolution will be consummated, there can be no doubt that the future belongs to aluminium, and that in times to come it will be the chief means of increasing human performance. It has in this respect capacities greater by far than those of any other metal. I should estimate its civilizing potency at fully one hundred times that of iron. This estimate, though it may astonish, is not at all exaggerated. First of all, we must remember that there is thirty times as much aluminium as iron in bulk, available for the uses of man. This in itself offers great possibilities. Then, again, the new metal is much more easily workable, which adds to its value. In many of its properties it partakes of the character of a precious metal, which gives it additional worth. Its electric conductivity, which, for a given weight, is greater than that of any other metal, would be alone sufficient to make it one of the most important factors in future human progress. Its extreme lightness makes it far more easy to transport the objects manufactured. By virtue of this property it will revolutionize naval construction, and in facilitating transport and travel it will add enormously to the useful performance of mankind. But its greatest civilizing potency will be, I believe, in aërial travel, which is sure to be brought about by means of it. Telegraphic instruments will slowly enlighten the barbarian. Electric motors and lamps will do it more quickly, but quicker than anything else the flying-machine will do it. By rendering travel ideally easy it will be the best means for unifying the heterogeneous elements of humanity. As the first step toward this realization we should produce a lighter storage-battery or get more energy from coal.

Efforts Toward Obtaining More Energy from Coal—The Electric Transmission—The Gas-Engine— The Cold-Coal Battery.

I remember that at one time I considered the production of electricity by burning coal in a battery as the greatest achievement toward advancing civilization, and I am surprised to find how much the continuous study of these subjects has modified

my views. It now seems to me that to burn coal, however efficiently, in a battery would be a mere makeshift, a phase in the
evolution toward something much more perfect. After all, in
generating electricity in this manner, we should be destroying
material, and this would be a barbarous process. We ought to
be able to obtain the energy we need without consumption of
material. But I am far from underrating the value of such an
efficient method of burning fuel. At the present time most motive
power comes from coal, and, either directly or by its products,
it adds vastly to human energy. Unfortunately, in all the processes now adopted, the larger portion of the energy of the coal
is uselessly dissipated. The best steam-engines utilize only a
small part of the total energy. Even in gas-engines, in which,
particularly of late, better results are obtainable, there is still a
barbarous waste going on. In our electric-lighting systems we
scarcely utilize one third of one per cent., and in lighting by
gas a much smaller fraction, of the total energy of the coal.
Considering the various uses of coal throughout the world, we
certainly do not utilize more than two per cent. of its energy
theoretically available. The man who should stop this senseless waste would be a great benefactor of humanity, though
the solution he would offer could not be a permanent one,
since it would ultimately lead to the exhaustion of the store of
material. Efforts toward obtaining more energy from coal are
now being made chiefly in two directions—by generating electricity and by producing gas for motive-power purposes. In
both of these lines notable success has already been achieved.

The advent of the alternating-current system of electric
power-transmission marks an epoch in the economy of energy
available to man from coal. Evidently all electrical energy
obtained from a waterfall, saving so much fuel, is a net gain to
mankind, which is all the more effective as it is secured with
little expenditure of human effort, and as this most perfect of
all known methods of deriving energy from the sun contributes
in many ways to the advancement of civilization. But electricity
enables us also to get from coal much more energy than was
practicable in the old ways. Instead of transporting the coal to
distant places of consumption, we burn it near the mine, develop

electricity in the dynamos, and transmit the current to remote localities, thus effecting a considerable saving. Instead of driving the machinery in a factory in the old wasteful way by belts and shafting, we generate electricity by steam-power and operate electric motors. In this manner it is not uncommon to obtain two or three times as much effective motive power from the fuel, besides securing many other important advantages. It is in this field as much as in the transmission of energy to great distances that the alternating system, with its ideally simple machinery, is bringing about an industrial revolution. But in many lines this progress has not yet been felt. For example, steamers and trains are still being propelled by the direct application of steam-power to shafts or axles. A much greater percentage of the heat-energy of the fuel could be transformed in motive energy by using, in place of the adopted marine engines and locomotives, dynamos driven by specially designed high-pressure steam- or gas-engines and by utilizing the electricity generated for the propulsion. A gain of fifty to one hundred per cent. in the effective energy derived from the coal could be secured in this manner. It is difficult to understand why a fact so plain and obvious is not receiving more attention from engineers. In ocean steamers such an improvement would be particularly desirable, as it would do away with noise and increase materially the speed and the carrying capacity of the liners.

Still more energy is now being obtained from coal by the latest improved gas-engine, the economy of which is, on the average, probably twice that of the best steam-engine. The introduction of the gas-engine is very much facilitated by the importance of the gas industry. With the increasing use of the electric light more and more of the gas is utilized for heating and motive-power purposes. In many instances gas is manufactured close to the coal-mine and conveyed to distant places of consumption, a considerable saving both in the cost of transportation and in utilization of the energy of the fuel being thus effected. In the present state of the mechanical and electrical arts the most rational way of deriving energy from coal is evidently to manufacture gas close to the coal store, and to utilize it, either on the spot or elsewhere, to generate electricity for industrial uses in dynamos driven by gas-engines. The commercial success of such a plant is largely dependent upon the production of gas-engines of great nominal horsepower, which, judging from the keen activity in this field, will soon be forthcoming. Instead of consuming coal directly, as usual, gas should be manufactured from it and burned to economize energy.

But all such improvements cannot be more than passing phases in the evolution toward something far more perfect, for ultimately we must succeed in obtaining electricity from coal in a more direct way, involving no great loss of its heat-energy. Whether coal can be oxidized by a cold process is still a question. Its combination with oxygen always evolves heat, and whether the energy of the combination of the carbon with another element can be turned directly into electrical energy has not yet been determined. Under certain conditions nitric acid will burn the carbon, generating an electric current, but the solution does not remain cold. Other means of oxidizing coal have been proposed, but they have offered no promise of leading to an efficient process. My own lack of success has been complete, though perhaps not quite so complete as that of some who have "perfected" the cold-coal battery. This problem is essentially one for the chemist to solve. It is not for the physicist, who determines all his results in advance, so that, when the experiment is tried, it cannot fail. Chemistry, though a positive science, does not yet admit of a solution by such positive methods as those which are available in the treatment of many physical problems. The result, if possible, will be arrived at through patient trying rather than through deduction or calculation. The time will soon come, however, when the chemist will be able to follow a course clearly mapped out beforehand, and when the process of his arriving at a desired result will be purely constructive. The cold-coal battery would give a great impetus to electrical development; it would lead very shortly to a practical flying-machine, and would enormously enhance the introduction of the automobile. But these and many other problems will be better solved, and in a more scientific manner, by a light-storage battery.

Energy from the Medium—
The Windmill and the Solar
Engine—Motive Power from
Terrestrial Heat—Electricity
from Natural Sources.

Besides fuel, there is abundant material from which we might eventually derive power. An immense amount of energy is locked up in limestone, for instance, and machines can be driven by liberating the carbonic acid through sulphuric acid or otherwise. I once constructed such an engine, and it operated satisfactorily.

But, whatever our resources of primary energy may be in the future, we must, to be rational, obtain it without consumption of any material. Long ago I came to this conclusion, and to arrive at this result only two ways, as before indicated, appeared possible—either to turn to use the energy of the sun stored in the ambient medium, or to transmit, through the medium, the sun's energy to distant places from some locality where it was obtainable without consumption of material. At that time I at once rejected the latter method as entirely impracticable, and turned to examine the possibilities of the former.

It is difficult to believe, but it is, nevertheless, a fact, that since time immemorial man has had at his disposal a fairly good machine which has enabled him to utilize the energy of the ambient medium. This machine is the windmill. Contrary to popular belief, the power obtainable from wind is very considerable. Many a deluded inventor has spent years of his life in endeavoring to "harness the tides," and some have even proposed to compress air by tide- or wave-power for supplying energy, never understanding the signs of the old windmill on the hill, as it sorrowfully waved its arms about and bade them stop. The fact is that a wave- or tide-motor would have, as a rule, but a small chance of competing commercially with the windmill, which is by far the better machine, allowing a much greater amount of energy to be obtained in a simpler way. Wind-power has been, in old times, of inestimable value to man, if for nothing else but for enabling him to cross the seas, and it is even now a very important factor in travel and transportation. But there are great limitations in this ideally simple method of utilizing the sun's energy. The machines are large for a given output, and the power is intermittent, thus necessitating the storage of energy and increasing the cost of the plant.

A far better way, however, to obtain power would be to avail ourselves of the sun's rays, which beat the earth incessantly and supply energy at a maximum rate of over four million horsepower per square mile. Although the average energy received per square mile in any locality during the year is only a small fraction of that amount, yet an inexhaustible source of power would be opened up by the discovery of some efficient method of utilizing the energy of the rays. The only rational way known to me at the time when I began the study of this subject was to employ some kind of heat- or thermodynamic engine, driven by a volatile fluid evaporated in a boiler by the heat of the rays. But closer investigation of this method, and calculation, showed that, notwithstanding the apparently vast amount of energy received from the sun's rays, only a small fraction of that energy could be actually utilized in this manner. Furthermore, the energy supplied through the sun's radiations is periodical, and the same limitations as in the use of the windmill I found to exist here also. After a long study of this mode of obtaining motive power from the sun, taking into account the necessarily large bulk of the boiler, the low efficiency of the heat-engine, the additional cost of storing the energy, and other drawbacks, I came to the conclusion that the "solar engine," a few instances excepted, could not be industrially exploited with success.

Another way of getting motive power from the medium without consuming any material would be to utilize the heat contained in the earth, the water, or the air for driving an engine. It is a well-known fact that the interior portions of the globe are very hot, the temperature rising, as observations show, with the approach to the center at the rate of approximately 1° C. for every hundred feet of depth. The difficulties of sinking shafts and placing boilers at depths of, say, twelve thousand feet, corresponding to an increase in temperature of about 120°C., are not insuperable, and we could certainly avail ourselves in this way of the internal heat of the globe. In fact, it would not be necessary to go to any depth at all in order to derive energy from the stored terrestrial heat. The superficial layers of the earth and the air strata close to the same are at a temperature sufficiently high to evaporate some extremely volatile substances, which we might use in our boilers instead of water. There is no doubt that a vessel might be propelled on the ocean by an engine driven by such a volatile fluid, no other energy being used but the heat abstracted from the water. But the amount of power which could be obtained in this manner would be, without further provision, very small.

Electricity produced by natural causes is another source of energy which might be rendered available. Lightning discharges involve great amounts of electrical energy, which we could utilize by transforming and storing it. Some years ago I made known a method of electrical transformation which renders the first part of this task easy, but the storing of the energy of lightning discharges will be difficult to accomplish. It is well known, furthermore, that electric currents circulate constantly through the earth, and that there exists between the earth and any air stratum a difference of electrical pressure, which varies in proportion to the height.

In recent experiments I have discovered two novel facts of importance in this connection. One of these facts is that an electric current is generated in a wire extending from the ground to a great height by the axial, and probably also by the translatory, movement of the earth. No appreciable current, however, will flow continuously in the wire unless the electricity is allowed to leak out into the air. Its escape is greatly facilitated by providing at the elevated end of the wire a conducting terminal of great surface, with many sharp edges or points. We are thus enabled to get a continuous supply of electrical energy by merely supporting a wire at a height, but, unfortunately, the amount of electricity which can be so obtained is small.

The second fact which I have ascertained is that the upper air strata are permanently charged with electricity opposite to that of the earth. So, at least, I have interpreted my observations, from which it appears that the earth, with its adjacent insulating and outer conducting envelop, constitutes a highly charged electrical condenser containing, in all probability, a great amount of electrical energy which might be turned to the uses of man, if it were possible to reach with a wire to great altitudes.

It is possible, and even probable, that there will be, in time, other resources of energy opened up, of which we have no knowledge now. We may even find ways of applying forces such as magnetism or gravity for driving machinery without using any other means. Such realizations, though highly improbable, are not impossible. An example will best convey an idea of what we can hope to attain and what we can never attain. Imagine a disk of some homogeneous material turned perfectly true and arranged to turn in frictionless bearings on a horizontal shaft above the ground. This disk, being under the above conditions perfectly balanced, would rest in any position. Now, it is possible that we may learn how to make such a disk rotate continuously and perform work by the force of gravity without any further effort on our part; but it is perfectly impossible for the disk to turn and to do work without any force from the outside. If it could do so, it would be what is designated scientifically as a "perpetuum mobile," a machine creating its own motive power. To make the disk rotate by the force of gravity we have only to invent a screen against this force. By such a screen we could prevent this force from acting on one half of the disk, and the rotation of the latter would follow. At least, we cannot deny such a possibility until we know exactly the nature of the force of gravity. Suppose that this force were due to a movement comparable to that of a stream of air passing from above toward the center of the earth. The effect of such a stream upon both halves of the disk would be equal, and the latter would not rotate ordinarily; but if one half should be guarded by a plate arresting the movement, then it would turn.

*A Departure from Known
Methods—Possibility of a "Self-
Acting" Engine or Machine,
Inanimate, Yet Capable, Like a
Living Being, of Deriving Energy
from the Medium—The Ideal Way of
Obtaining Motive Power.*

When I began the investigation of the subject under consideration, and when the preceding or similar ideas presented themselves to me for the first time, though I was then unacquainted
with a number of the facts mentioned, a survey of the various
ways of utilizing the energy of the medium convinced me, nevertheless, that to arrive at a thoroughly satisfactory practical
solution a radical departure from the methods then known
had to be made. The windmill, the solar engine, the engine
driven by terrestrial heat, had their limitations in the amount
of power obtainable. Some new way had to be discovered which
would enable us to get more energy. There was enough heat-
energy in the medium, but only a small part of it was available
for the operation of an engine in the ways then known. Besides,
the energy was obtainable only at a very slow rate. Clearly, then,
the problem was to discover some new method which would
make it possible both to utilize more of the heat-energy of the
medium and also to draw it away from the same at a more
rapid rate.

I was vainly endeavoring to form an idea of how this might
be accomplished, when I read some statements from Carnot
and Lord Kelvin (then Sir William Thomson) which meant virtually that it is impossible for an inanimate mechanism or
self-acting machine to cool a portion of the medium below
the temperature of the surrounding, and operate by the heat
abstracted. These statements interested me intensely. Evidently
a living being could do this very thing, and since the experiences of my early life which I have related had convinced me
that a living being is only an automaton, or, otherwise stated,
a "self-acting engine," I came to the conclusion that it was
possible to construct a machine which would do the same. As

the first step toward this realization I conceived the following mechanism. Imagine a thermopile consisting of a number of bars of metal extending from the earth to the outer space beyond the atmosphere. The heat from below, conducted upward along these metal bars, would cool the earth or the sea or the air, according to the location of the lower parts of the bars, and the result, as is well known, would be an electric current circulating in these bars. The two terminals of the thermopile could now be joined through an electric motor, and, theoretically, this motor would run on and on, until the media below would be cooled down to the temperature of the outer space. This would be an inanimate engine which, to all evidence, would be cooling a portion of the medium below the temperature of the surrounding, and operating by the heat abstracted.

But was it not possible to realize a similar condition without necessarily going to a height? Conceive, for the sake of illustration, an inclosure T, as illustrated in diagram b, such that energy could not be transferred across it except through a channel or path O, and that, by some means or other, in this inclosure a medium were maintained which would have little energy, and that on the outer side of the same there would be the ordinary ambient medium with much energy. Under these assumptions the energy would flow through the path O, as indicated by the arrow, and might then be converted on its passage into some other form of energy. The question was, Could such a condition be attained? Could we produce artificially such a "sink" for the energy of the ambient medium to flow in? Suppose that an extremely low temperature could be maintained by some process in a given space; the surrounding medium would then be compelled to give off heat, which could be converted into mechanical or other form of energy, and utilized. By realizing such a plan, we should be enabled to get at any point of the globe a continuous supply of energy, day and night. More than this, reasoning in the abstract, it would seem possible to cause a quick circulation of the medium, and thus draw the energy at a very rapid rate.

Here, then, was an idea which, if realizable, afforded a happy solution of the problem of getting energy from the medium.

But was it realizable? I convinced myself that it was so in a number of ways, of which one is the following. As regards heat, we are at a high level, which may be represented by the surface of a mountain lake considerably above the sea, the level of which may mark the absolute zero of temperature existing in the interstellar space. Heat, like water, flows from high to low level, and, consequently, just as we can let the water of the lake run down to the sea, so we are able to let heat from the earth's surface travel up into the cold region above. Heat, like water, can perform work in flowing down, and if we had any doubt as to whether we could derive energy from the medium by means of a thermopile, as before described, it would be dispelled by this analogue. But can we produce cold in a given portion of the space and cause the heat to flow in continually? To create such a "sink," or "cold hole," as we might say, in the medium, would be equivalent to producing in the lake a space either empty or filled with something much lighter than water. This we could do by placing in the lake a tank, and pumping all the water out of the latter. We know, then, that the water, if allowed to flow back into the tank, would, theoretically, be able to perform exactly the same amount of work which was used in pumping it out, but not a bit more. Consequently nothing could be gained in this double operation of first raising the water and then letting it fall down. This would mean that it is impossible to create such a sink in the medium. But let us reflect a moment. Heat, though following certain general laws of mechanics, like a fluid, is not such; it is energy which may be converted into other forms of energy as it passes from a high to a low level. To make our mechanical analogy complete and true, we must, therefore, assume that the water, in its passage into the tank, is converted into something else, which may be taken out of it without using any, or by using very little, power. For example, if heat be represented in this analogue by the water of the lake, the oxygen and hydrogen composing the water may illustrate other forms of energy into which the heat is transformed in passing from hot to cold. If the process of heat-transformation were absolutely perfect, no heat at all would arrive at the low level, since all of it would be converted into other forms of energy. Corresponding to this ideal case, all the water flowing into the tank would be decomposed into oxygen and hydrogen before reaching the bottom, and the result would be that water would continually flow in, and yet the tank would remain entirely empty, the gases formed escaping. We would thus produce, by expending initially a certain amount of work to create a sink for the heat or, respectively, the water to flow in, a condition enabling us to get any amount of energy without further effort. This would be an ideal way of obtaining motive power. We do not know of any such absolutely perfect process of heat-conversion, and consequently some heat will generally reach the low level, which means to say, in our mechanical analogue, that some water will arrive at the bottom of the tank, and a gradual and slow filling of the latter will take place, necessitating continuous pumping out. But evidently there will be less to pump out than flows in, or, in other words, less energy will be needed to maintain the initial condition than is developed by the fall, and this is to say that some energy will be gained from the medium. What is not converted in flowing down can just be raised up with its own energy, and what is converted is clear gain. Thus the virtue of the principle I have discovered resides wholly in the conversion of the energy on the downward flow.

First Efforts to Produce the Self-Acting Engine—The Mechanical Oscillator—Work of Dewar and Linde—Liquid Air.

Having recognized this truth, I began to devise means for carrying out my idea, and, after long thought, I finally conceived a combination of apparatus which should make possible the obtaining of power from the medium by a process of continuous cooling of atmospheric air. This apparatus, by continually transforming heat into mechanical work, tended to become colder and colder, and if it only were practicable to reach a very low temperature in this manner, then a sink for the heat could be produced, and energy could be derived from the medium. This seemed to be contrary to the statements of Carnot and Lord Kelvin before referred to, but I concluded from the theory of the process that such a result could be attained. This conclusion I reached, I think, in the latter part of 1883, when I was in Paris, and it was at a time when my mind was being more and more dominated by an invention which I had evolved during the preceding year, and which has since become known under the name of the "rotating magnetic field." During the few years which followed I elaborated further the plan I had imagined, and studied the working conditions, but made little headway. The commercial introduction in this country of the invention before referred to required most of my energies until 1889, when I again took up the idea of the self-acting machine. A closer investigation of the principles involved, and calculation, now showed that the result I aimed at could not be reached in a practical manner by ordinary machinery, as I had in the beginning expected. This led me, as a next step, to the study of a type of engine generally designated as "turbine," which at first seemed to offer better chances for a realization of the idea. Soon I found, however, that the turbine, too, was unsuitable. But my conclusions showed that if an engine of a peculiar kind could be brought to a high degree of perfection, the plan I had conceived was realizable, and I resolved to proceed with the development of such an engine, the primary object of which was to secure the greatest economy of transformation of heat into mechanical energy. A characteristic feature of the engine was that the work-performing piston was not connected with anything else, but was perfectly free to vibrate at an enormous rate. The mechanical difficulties encountered in the construction of this engine were greater than I had anticipated, and I made slow progress. This work was continued until early in 1892, when I went to London, where I saw Professor Dewar's admirable experiments with liquefied gases. Others had liquefied gases before, and notably Ozlewski and Pictet had performed creditable early experiments in this line, but there was such a vigor about the work of Dewar that even the old appeared new. His experiments showed, though in a way different from that I had imagined, that it was possible to reach a very low temperature by transforming heat into mechanical work, and I returned, deeply impressed with what I had seen, and more than ever convinced that my plan was practicable. The work temporarily interrupted was taken up anew, and soon I had in a fair state of perfection the engine which I have named "the mechanical oscillator." In this machine I succeeded in doing away with all packings, valves, and lubrication, and in producing so rapid a vibration of the piston that shafts of tough steel, fastened to the same and vibrated longitudinally, were torn asunder. By combining this engine with a dynamo of special design I produced a highly efficient electrical generator, invaluable in measurements and determinations of physical quantities on account of the unvarying rate of oscillation obtainable by its means. I exhibited several types of this machine, named "mechanical and electrical oscillator," before the Electrical Congress at the World's Fair in Chicago during the summer of 1893, in a lecture which, on account of other pressing work, I was unable to prepare for publication. On that occasion I exposed the principles of the mechanical oscillator, but the original purpose of this machine is explained here for the first time.

In the process, as I had primarily conceived it, for the utilization of the energy of the ambient medium, there were five essential elements in combination, and each of these had to be newly designed and perfected, as no such machines existed. The mechanical oscillator was the first element of this combination, and having perfected this, I turned to the next, which was an air-compressor of a design in certain respects resembling that of the mechanical oscillator. Similar difficulties in the construction were again encountered, but the work was pushed vigorously, and at the close of 1894 I had completed these two elements of the combination, and thus produced an apparatus for compressing air, virtually to any desired pressure, incomparably simpler, smaller, and more efficient than the ordinary. I was just beginning work on the third element, which together with the first two would give a refrigerating machine of exceptional efficiency and simplicity, when a misfortune befell me in the burning of my laboratory, which crippled my labors and delayed me. Shortly afterward Dr. Carl Linde announced the liquefaction of air by a self-cooling process, demonstrating that it was practicable to proceed with the cooling until liquefaction of the air took place. This was the only experimental proof which I was still wanting that energy was obtainable from the medium in the manner contemplated by me.

The liquefaction of air by a self-cooling process was not, as popularly believed, an accidental discovery, but a scientific result which could not have been delayed much longer, and which, in all probability, could not have escaped Dewar. This fascinating advance, I believe, is largely due to the powerful work of this great Scotchman. Nevertheless, Linde's is an immortal achievement. The manufacture of liquid air has been carried on for four years in Germany, on a scale much larger than in any other country, and this strange product has been applied for a variety of purposes. Much was expected of it in the beginning, but so far it has been an industrial ignis fatuus. By the use of such machinery as I am perfecting, its cost will probably be greatly lessened, but even then its commercial success will be questionable. When used as a refrigerant it is uneconomical, as its temperature is unnecessarily low. It is as expensive to maintain a body at a very low temperature as it is to keep it very hot; it takes coal to keep air cold. In oxygen manufacture it cannot yet compete with the electrolytic method. For use as an explosive it is unsuitable, because its low temperature again condemns it to a small efficiency, and for motive-power purposes its cost is still by far too high. It is of interest to note, however, that in driving an engine by liquid air a certain amount of energy may be gained from the engine, or, stated otherwise, from the ambient medium which keeps the engine warm, each two hundred pounds of iron-casting of the latter contributing energy at the rate of about one effective horse-power during one hour. But this gain of the consumer is offset by an equal loss of the producer.

Much of this task on which I have labored so long remains to be done. A number of mechanical details are still to be perfected and some difficulties of a different nature to be mastered, and I cannot hope to produce a self-acting machine deriving energy from the ambient medium for a long time yet, even if all my expectations should materialize. Many circumstances have occurred which have retarded my work of late, but for several reasons the delay was beneficial.

One of these reasons was that I had ample time to consider what the ultimate possibilities of this development might be. I worked for a long time fully convinced that the practical realization of this method of obtaining energy from the sun would be of incalculable industrial value, but the continued study of the subject revealed the fact that while it will be commercially profitable if my expectations are well founded, it will not be so to an extraordinary degree.

*Discovery of Unexpected
Properties of the Atmosphere—
Strange Experiments—
Transmission of Electrical Energy
Through One Wire Without
Return—Transmission Through the
Earth Without Any Wire.*

Another of these reasons was that I was led to recognize the transmission of electrical energy to any distance through the media as by far the best solution of the great problem of harnessing the sun's energy for the uses of man. For a long time I was convinced that such a transmission on an industrial scale could never be realized, but a discovery which I made changed my view. I observed that under certain conditions the atmosphere, which is normally a high insulator, assumes conducting properties, and so becomes capable of conveying any amount of electrical energy. But the difficulties in the way of a practical utilization of this discovery for the purpose of transmitting electrical energy without wires were seemingly insuperable. Electrical pressures of many millions of volts had to be produced and handled; generating apparatus of a novel kind, capable of withstanding the immense electrical stresses, had to be invented and perfected, and a complete safety against the dangers of the high-tension currents had to be attained in the system before its practical introduction could be even thought of. All this could not be done in a few weeks or months, or even years. The work required patience and constant application, but the improvements came, though slowly. Other valuable results were, however, arrived at in the course of this long-continued work, of which I shall endeavor to give a brief account, enumerating the chief advances as they were successively effected.

The discovery of the conducting properties of the air, though unexpected, was only a natural result of experiments in a special field which I had carried on for some years before. It was, I believe, during 1889 that certain possibilities offered by extremely rapid electrical oscillations determined me to design a number of special machines adapted for their investigation. Owing to the peculiar requirements, the construction of these machines was very difficult, and consumed much time and effort; but my work on them was generously rewarded, for I reached by their means several novel and important results. One of the earliest observations I made with these new machines was that electrical oscillations of an extremely high rate act in an extraordinary manner upon the human organism. Thus, for instance, I demonstrated that powerful electrical discharges of several hundred thousand volts, which at that time were considered absolutely deadly, could be passed through the body without inconvenience or hurtful consequences. These oscillations produced other specific physiological effects, which, upon my announcement, were eagerly taken up by skilled physicians and further investigated. This new field has proved itself fruitful beyond expectation, and in the few years which have passed since, it has been developed to such an extent that it now forms a legitimate and important department of medical science. Many results, thought impossible at that time, are now readily obtainable with these oscillations, and many experiments undreamed of then can now be readily performed by their means. I still remember with pleasure how, nine years ago, I passed the discharge of a powerful induction-coil through my body to demonstrate before a scientific society the comparative harmlessness of very rapidly vibrating electric currents, and I can still recall the astonishment of my audience. I would now undertake, with much less apprehension than I had in that experiment, to transmit through my body with such currents the entire electrical energy of the dynamos now working at Niagara—forty or fifty thousand horse-power. I have produced electrical oscillations which were of such intensity that when circulating through my arms and chest they have melted wires which joined my hands, and still I felt no inconvenience. I have energized with such oscillations a loop of heavy copper wire so powerfully that masses of metal, and even objects of an electrical resistance specifically greater than that of human tissue, brought close to or placed within the loop, were heated to a high temperature and melted, often with the violence of an explosion, and yet into this very space in which this terribly destructive turmoil was going on I have repeatedly thrust my head without feeling anything or experiencing injurious after-effects.

Another observation was that by means of such oscillations light could be produced in a novel and more economical manner, which promised to lead to an ideal system of electric illumination by vacuum-tubes, dispensing with the necessity of renewal of lamps or incandescent filaments, and possibly also with the use of wires in the interior of buildings. The efficiency of this light increases in proportion to the rate of the oscillations, and its commercial success is, therefore, dependent on the economical production of electrical vibrations of transcending rates. In this direction I have met with gratifying success of late, and the practical introduction of this new system of illumination is not far off.

The investigations led to many other valuable observations and results, one of the more important of which was the demonstration of the practicability of supplying electrical energy through one wire without return. At first I was able to transmit in this novel manner only very small amounts of electrical energy, but in this line also my efforts have been rewarded with similar success.

The photograph shown in Fig.3 illustrates, as its title explains, an actual transmission of this kind effected with apparatus used in other experiments here described. To what a degree the appliances have been perfected since my first demonstrations early in 1891 before a scientific society, when my apparatus was barely capable of lighting one lamp (which result was considered wonderful), will appear when I state that I have now no difficulty in lighting in this manner four or five hundred lamps, and could light many more. In fact, there is no limit to the amount of energy which may in this way be supplied to operate any kind of electrical device.

After demonstrating the practicability of this method of transmission, the thought naturally occurred to me to use the earth as a conductor, thus dispensing with all wires. Whatever electricity may be, it is a fact that it behaves like an incompressible fluid, and the earth may be looked upon as an immense reservoir of electricity, which, I thought, could be disturbed effectively by a properly designed electrical machine. Accordingly, my next efforts were directed toward perfecting a special apparatus which would be highly effective in creating a disturbance of electricity in the earth. The progress in this new direction was necessarily very slow and the work discouraging, until I finally succeeded in perfecting a novel kind of transformer or induction-coil, particularly suited for this special purpose. That it is practicable, in this manner, not only to transmit minute amounts of electrical energy for operating delicate electrical devices, as I contemplated at first, but also electrical energy in appreciable quantities, will appear from an inspection of Fig. 4, which illustrates an actual experiment of this kind performed with the same apparatus. The result obtained was all the more remarkable as the top end of the coil was not connected to a wire or plate for magnifying the effect.

"Wireless" Telegraphy—The Secret of Tuning—Errors in the Hertzian Investigations—A Receiver of Wonderful Sensitiveness.

As the first valuable result of my experiments in this latter line a system of telegraphy without wires resulted, which I described in two scientific lectures in February and March, 1893. It is mechanically illustrated in diagram c, the upper part of which shows the electrical arrangement as I described it then, while the lower part illustrates its mechanical analogue. The system is extremely simple in principle. Imagine two tuning-forks F, F_1, one at the sending- and the other at the receiving-station respectively, each having attached to its lower prong a minute piston p, fitting in a cylinder. Both the cylinders communicate with a large reservoir R, with elastic walls, which is supposed to be closed and filled with a light and incompressible fluid. By striking repeatedly one of the prongs of the tuning-fork F, the small piston p below would be vibrated, and its vibrations, transmitted through the fluid, would reach the distant fork F_1, which is "tuned" to the fork F, or, stated otherwise, of exactly the same note as the latter. The fork F_1 would now be set vibrating, and its vibration would be intensified by the continued action of the distant fork F until its upper prong, swinging far out, would make an electrical connection with a stationary contact c'', starting in this manner some electrical or other appliances which may be used for recording the signals. In this simple way messages could be exchanged between the two stations, a similar contact c' being provided for this purpose, close to the upper prong of the fork F, so that the apparatus at each station could be employed in turn as receiver and transmitter.

The electrical system illustrated in the upper figure of diagram c is exactly the same in principle, the two wires or circuits ESP and $E_1S_1P_1$, which extend vertically to a height, representing the two tuning-forks with the pistons attached to them. These circuits are connected with the ground by plates E, E_1, and to two elevated metal sheets P, P_1, which store electricity and thus magnify considerably the effect. The closed reservoir R, with elastic walls, is in this case replaced by the earth, and the fluid by electricity. Both of these circuits are "tuned" and operate just like the two tuning-forks. Instead of striking the fork F at the sending-station, electrical oscillations are produced in the vertical sending- or transmitting-wire ESP, as by the action of a source S, included in this wire, which spread through the ground and reach the distant vertical receiving-wire

$E_I S_I P_I$, exciting corresponding electrical oscillations in the same. In the latter wire or circuit is included a sensitive device or receiver S_I, which is thus set in action and made to operate a relay or other appliance. Each station is, of course, provided both with a source of electrical oscillations S and a sensitive receiver S_I, and a simple provision is made for using each of the two wires alternately to send and to receive the messages.

The exact attunement of the two circuits secures great advantages, and, in fact, it is essential in the practical use of the system. In this respect many popular errors exist, and, as a rule, in the technical reports on this subject circuits and appliances are described as affording these advantages when from their very nature it is evident that this is impossible. In order to attain the best results it is essential that the length of each wire or circuit, from the ground connection to the top, should be equal to one quarter of the wave-length of the electrical vibration in the wire, or else equal to that length multiplied by an odd number. Without the observation of this rule it is virtually impossible to prevent the interference and insure the privacy of messages. Therein lies the secret of tuning. To obtain the most satisfactory results it is, however, necessary to resort to electrical vibrations of low pitch. The Hertzian spark apparatus, used generally by experimenters, which produces oscillations of a very high rate, permits no effective tuning, and slight disturbances are sufficient to render an exchange of messages impracticable. But scientifically designed, efficient appliances allow nearly perfect adjustment. An experiment performed with the improved apparatus repeatedly referred to, and intended to convey an idea of this feature, is illustrated in Fig. 5, which is sufficiently explained by its note.

Since I described these simple principles of telegraphy without wires I have had frequent occasion to note that the identical features and elements have been used, in the evident belief that the signals are being transmitted to considerable distances by "Hertzian" radiations. This is only one of many misapprehensions to which the investigations of the lamented physicist have given rise. About thirty-three years ago Maxwell, following up a suggestive experiment made by Faraday in 1845, evolved an ideally simple theory which intimately connected light, radiant heat, and electrical phenomena, interpreting them as being all due to vibrations of a hypothetical fluid of inconceivable tenuity, called the ether. No experimental verification was arrived at until Hertz, at the suggestion of Helmholtz, undertook a series of experiments to this effect. Hertz proceeded with extraordinary ingenuity and insight, but devoted little energy to the perfection of his old-fashioned apparatus. The consequence was that he failed to observe the important function which the air played in his experiments, and which I subsequently discovered. Repeating his experiments and reaching different results, I ventured to point out this oversight. The strength of the proofs brought forward by Hertz in support of Maxwell's theory resided in the correct estimate of the rates of vibration of the circuits he used. But I ascertained that he could not have obtained the rates he thought he was getting. The vibrations with identical apparatus he employed are, as a rule, much slower, this being due to the presence of air, which produces a dampening effect upon a rapidly vibrating electric circuit of high pressure, as a fluid does upon a vibrating tuning-fork. I have, however, discovered since that time other causes of error, and I have long ago ceased to look upon his results as being an experimental verification of the poetical conceptions of Maxwell. The work of the great German physicist has acted as an immense stimulus to contemporary electrical research, but it has likewise, in a measure, by its fascination, paralyzed the scientific mind, and thus hampered independent inquiry. Every new phenomenon which was discovered was made to fit the theory, and so very often the truth has been unconsciously distorted.

When I advanced this system of telegraphy, my mind was dominated by the idea of effecting communication to any distance through the earth or environing medium, the practical consummation of which I considered of transcendent importance, chiefly on account of the moral effect which it could not fail to produce universally. As the first effort to this end I proposed, at that time, to employ relay-stations with tuned circuits, in the hope of making thus practicable signaling over vast distances, even with apparatus of very moderate power then at my command. I was confident, however, that with properly designed machinery signals could be transmitted to any point of the globe, no matter what the distance, without the necessity of using such intermediate stations. I gained this conviction through the discovery of a singular electrical phenomenon, which I described early in 1892, in lectures delivered before some scientific societies abroad, and which I have called a "rotating brush." This is a bundle of light which is formed, under certain conditions, in a vacuum-bulb, and which is of a sensitiveness to magnetic and electric influences bordering, so to speak, on the supernatural. This light-bundle is rapidly rotated by the earth's magnetism as many as twenty thousand times per second, the rotation in these parts being opposite to what it would be in the southern hemisphere, while in the region of the magnetic equator it should not rotate at all. In its most sensitive state, which is difficult to attain, it is responsive to electric or magnetic influences to an incredible degree. The mere stiffening of the muscles of the arm and consequent slight electrical change in the body of an observer standing at some distance from it, will perceptibly affect it. When in this highly sensitive state it is capable of indicating the slightest magnetic and electric changes taking place in the earth. The observation of this wonderful phenomenon impressed me strongly that communication at any distance could be easily effected by its means, provided that apparatus could be perfected capable of producing an electric or magnetic change of state, however small, in the terrestrial globe or environing medium.

Development of a New Principle—
The Electrical Oscillator—
Production of Immense Electrical
Movements—The Earth Responds
to Man—Interplanetary
Communication Now Probable.

I resolved to concentrate my efforts upon this venturesome task, though it involved great sacrifice, for the difficulties to be mastered were such that I could hope to consummate it only after years of labor. It meant delay of other work to which I would have preferred to devote myself, but I gained the conviction that my energies could not be more usefully employed; for I recognized that an efficient apparatus for the production of powerful electrical oscillations, as was needed for that specific purpose, was the key to the solution of other most important electrical and, in fact, human problems. Not only was communication, to any distance, without wires possible by its means, but, likewise, the transmission of energy in great amounts, the burning of the atmospheric nitrogen, the production of an efficient illuminant, and many other results of inestimable scientific and industrial value. Finally, however, I had the satisfaction of accomplishing the task undertaken by the use of a new principle, the virtue of which is based on the marvelous properties of the electrical condenser. One of these is that it can discharge or explode its stored energy in an inconceivably short time. Owing to this it is unequaled in explosive violence. The explosion of dynamite is only the breath of a consumptive compared with its discharge. It is the means of producing the strongest current, the highest electrical pressure, the greatest commotion in the medium. Another of its properties, equally valuable, is that its discharge may vibrate at any rate desired up to many millions per second.

I had arrived at the limit of rates obtainable in other ways when the happy idea presented itself to me to resort to the condenser. I arranged such an instrument so as to be charged and discharged alternately in rapid succession through a coil with a few turns of stout wire, forming the primary of a transformer or induction-coil. Each time the condenser was discharged the current would quiver in the primary wire and induce corresponding oscillations in the secondary. Thus a transformer or induction-coil on new principles was evolved, which I have called "the electrical oscillator," partaking of those unique qualities which characterize the condenser, and enabling results to be attained impossible by other means. Electrical effects of any desired character and of intensities undreamed of before are now easily producible by perfected apparatus of this kind, to which frequent reference has been made, and the essential parts of which are shown in Fig. 6. For certain purposes a strong inductive effect is required; for others the greatest possible suddenness; for others again, an exceptionally high rate of vibration or extreme pressure; while for certain other objects immense electrical movements are necessary. The photographs in Figs. 7, 8, 9, and 10, of experiments performed with such an oscillator, may serve to illustrate some of these features and convey an idea of the magnitude of the effects actually produced. The completeness of the titles of the figures referred to makes a further description of them unnecessary.

However extraordinary the results shown may appear, they are but trifling compared with those which are attainable by apparatus designed on these same principles. I have produced electrical discharges the actual path of which, from end to end, was probably more than one hundred feet long; but it would not be difficult to reach lengths one hundred times as great. I have produced electrical movements occurring at the rate of approximately one hundred thousand horse-power, but rates of one, five, or ten million horse-power are easily practicable. In these experiments effects were developed incomparably greater than any ever produced by human agencies, and yet these results are but an embryo of what is to be.

That communication without wires to any point of the globe is practicable with such apparatus would need no demonstration, but through a discovery which I made I obtained absolute certitude. Popularly explained, it is exactly this: When we raise the voice and hear an echo in reply, we know that the sound of the voice must have reached a distant wall, or boundary, and must have been reflected from the same. Exactly as the sound, so an electrical wave is reflected, and the same evidence which is afforded by an echo is offered by an electrical phenomenon known as a "stationary" wave—that is, a wave with fixed nodal and ventral regions. Instead of sending sound-vibrations toward a distant wall, I have sent electrical vibrations toward the remote boundaries of the earth, and instead of the wall the earth has replied. In place of an echo I have obtained a stationary electrical wave, a wave reflected from afar.

Stationary waves in the earth mean something more than mere telegraphy without wires to any distance. They will enable us to attain many important specific results impossible otherwise. For instance, by their use we may produce at will, from a sending-station, an electrical effect in any particular region of the globe; we may determine the relative position or course of a moving object, such as a vessel at sea, the distance traversed by the same, or its speed; or we may send over the earth a wave of electricity traveling at any rate we desire, from the pace of a turtle up to lightning speed.

With these developments we have every reason to anticipate that in a time not very distant most telegraphic messages across the oceans will be transmitted without cables. For short distances we need a "wireless" telephone, which requires no expert operators. The greater the spaces to be bridged, the more rational becomes communication without wires. The cable is not only an easily damaged and costly instrument, but it limits us in the speed of transmission by reason of a certain electrical property inseparable from its construction. A properly designed plant for effecting communication without wires ought to have many times the working capacity of a cable, while it will involve incomparably less expense. Not a long time will pass, I believe, before communication by cable will become obsolete, for not only will signaling by this new method be quicker and cheaper, but also much safer. By using some new means for isolating the messages which I have contrived, an almost perfect privacy can be secured.

I have observed the above effects so far only up to a limited distance of about six hundred miles, but inasmuch as there is virtually no limit to the power of the vibrations producible with such an oscillator, I feel quite confident of the success of such a plant for effecting transoceanic communication. Nor is this all. My measurements and calculations have shown that it is perfectly practicable to produce on our globe, by the use of these principles, an electrical movement of such magnitude that, without the slightest doubt, its effect will be perceptible on some of our nearer planets, as Venus and Mars. Thus from mere possibility interplanetary communication has entered the stage of probability. In fact, that we can produce a distinct effect on one of these planets in this novel manner, namely, by disturbing the electrical condition of the earth, is beyond any doubt. This way of effecting such communication is, however, essentially different from all others which have so far been proposed by scientific men. In all the previous instances only a minute fraction of the total energy reaching the planet—as much as it would be possible to concentrate in a reflector—could be utilized by the supposed observer in his instrument. But by the means I have developed he would be enabled to concentrate the larger portion of the entire energy transmitted to the planet in his instrument, and the chances of affecting the latter are thereby increased many millionfold.

Besides machinery for producing vibrations of the required power, we must have delicate means capable of revealing the effects of feeble influences exerted upon the earth. For such purposes, too, I have perfected new methods. By their use we shall likewise be able, among other things, to detect at considerable distance the presence of an iceberg or other object at sea. By their use, also, I have discovered some terrestrial phenomena still unexplained. That we can send a message to a planet is certain, that we can get an answer is probable: man is not the only being in the Infinite gifted with a mind.

Transmission of Electrical Energy to Any Distance Without Wires— Now Practicable—The Best Means of Increasing the Force Accelerating the Human Mass.

The most valuable observation made in the course of these investigations was the extraordinary behavior of the atmosphere toward electric impulses of excessive electromotive force. The experiments showed that the air at the ordinary pressure became distinctly conducting, and this opened up the wonderful prospect of transmitting large amounts of electrical energy for industrial purposes to great distances without wires, a possibility which, up to that time, was thought of only as a scientific dream. Further investigation revealed the important fact that the conductivity imparted to the air by these electrical impulses of many millions of volts increased very rapidly with the degree of rarefaction, so that air strata at very moderate altitudes, which are easily accessible, offer, to all experimental evidence, a perfect conducting path, better than a copper wire, for currents of this character.

Thus the discovery of these new properties of the atmosphere not only opened up the possibility of transmitting, without wires, energy in large amounts, but, what was still more significant, it afforded the certitude that energy could be transmitted in this manner economically. In this new system it matters little—in fact, almost nothing—whether the transmission is effected at a distance of a few miles or of a few thousand miles.

While I have not, as yet, actually effected a transmission of a considerable amount of energy, such as would be of industrial importance, to a great distance by this new method, I have operated several model plants under exactly the same conditions which will exist in a large plant of this kind, and the practicability of the system is thoroughly demonstrated. The experiments have shown conclusively that, with two terminals maintained at an elevation of not more than thirty thousand to thirty-five thousand feet above sea-level, and with an electrical pressure of fifteen to twenty million volts, the energy of thousands of horse-power can be transmitted over distances which may be hundreds and, if necessary, thousands of miles. I am hopeful, however, that I may be able to reduce very considerably the elevation of the terminals now required, and with this object I am following up an idea which promises such a realization. There is, of course, a popular prejudice against using an electrical pressure of millions of volts, which may cause sparks to fly at distances of hundreds of feet, but, paradoxical as it may seem, the system, as I have described it in a technical publication, offers greater personal safety than most of the ordinary distribution circuits now used in the cities. This is, in a measure, borne out by the fact that, although I have carried on such experiments for a number of years, no injury has been sustained either by me or any of my assistants.

But to enable a practical introduction of the system, a number of essential requirements are still to be fulfilled. It is not enough to develop appliances by means of which such a transmission can be effected. The machinery must be such as to allow the transformation and transmission of electrical energy under highly economical and practical conditions. Furthermore, an inducement must be offered to those who are engaged in the industrial exploitation of natural sources of power, as waterfalls, by guaranteeing greater returns on the capital invested than they can secure by local development of the property.

From that moment when it was observed that, contrary to the established opinion, low and easily accessible strata of the atmosphere are capable of conducting electricity, the transmission of electrical energy without wires has become a rational task of the engineer, and one surpassing all others in importance. Its practical consummation would mean that energy would be available for the uses of man at any point of the globe, not in small amounts such as might be derived from the ambient medium by suitable machinery, but in quantities virtually unlimited, from waterfalls. Export of power would then become the chief source of income for many happily situated countries, as the United States, Canada, Central and South America, Switzerland, and Sweden. Men could settle down everywhere, fertilize and irrigate the soil with little effort, and convert barren deserts into gardens, and thus the entire globe could be transformed and made a fitter abode for mankind. It is highly probable that if there are intelligent beings on Mars they have long ago realized this very idea, which would explain the changes on its surface noted by astronomers. The atmosphere on that planet, being of considerably smaller density than that of the earth, would make the task much more easy.

It is probable that we shall soon have a self-acting heat-engine capable of deriving moderate amounts of energy from the ambient medium. There is also a possibility—though a small one—that we may obtain electrical energy direct from the sun. This might be the case if the Maxwellian theory is true, according to which electrical vibrations of all rates should emanate from the sun. I am still investigating this subject. Sir William Crookes has shown in his beautiful invention known as the "radiometer" that rays may produce by impact a mechanical effect, and this may lead to some important revelation as to the utilization of the sun's rays in novel ways. Other sources of energy may be opened up, and new methods of deriving energy from the sun discovered, but none of these or similar achievements would equal in importance the transmission of power to any distance through the medium. I can conceive of no technical advance which would tend to unite the various elements of humanity more effectively than this one, or of one which would more add to and more economize human energy. It would be the best means of increasing the force accelerating the human mass. The mere moral influence of such a radical departure would be incalculable. On the other hand, if at any point of the globe energy can be obtained in limited quantities from the ambient medium by means of a self-acting heat-engine or otherwise, the conditions will remain the same as before. Human performance will be increased, but men will remain strangers as they were.

I anticipate that many, unprepared for these results, which, through long familiarity, appear to me simple and obvious, will consider them still far from practical application. Such reserve, and even opposition, of some is as useful a quality and as necessary an element in human progress as the quick receptivity and enthusiasm of others. Thus, a mass which resists the force at first, once set in movement, adds to the energy. The scientific man does not aim at an immediate result. He does not expect that his advanced ideas will be readily taken up. His work is like that of the planter—for the future. His duty is to lay the foundation for those who are to come, and point the way. He lives and labors and hopes with the poet who says:

> Schaff', das Tagwerk meiner Hände,
> Hohes Glück, dass ich's vollende!
> Lass, o lass mich nicht ermatten!
> Nein, es sind nicht leere Träume:
> Jetzt nur Stangen, diese Bäume
> Geben einst noch Frucht und Schatten.*

---

* Daily work—my hands' employment,
  To complete is pure enjoyment!
  Let, oh, let me never falter!
  No! there is no empty dreaming:
  Lo! these trees, but bare poles seeming,
  Yet will yield both fruit and shelter!
    GOETHE'S "HOPE,"
    Translated by William Gibson, Com. U. S. N.

THE STORY OF PENGUIN CLASSICS

Before 1946 . . . "Classics" are mainly the domain of academics and students; readable editions for everyone else are almost unheard of. This all changes when a little-known classicist, E. V. Rieu, presents Penguin founder Allen Lane with the translation of Homer's *Odyssey* that he has been working on in his spare time.

1946 Penguin Classics debuts with *The Odyssey,* which promptly sells three million copies. Suddenly, classics are no longer for the privileged few.

1950s Rieu, now series editor, turns to professional writers for the best modern, readable translations, including Dorothy L. Sayers's *Inferno* and Robert Graves's unexpurgated *Twelve Caesars.*

1960s The Classics are given the distinctive black covers that have remained a constant throughout the life of the series. Rieu retires in 1964, hailing the Penguin Classics list as "the greatest educative force of the twentieth century."

1970s A new generation of translators swells the Penguin Classics ranks, introducing readers of English to classics of world literature from more than twenty languages. The list grows to encompass more history, philosophy, science, religion, and politics.

1980s The Penguin American Library launches with titles such as *Uncle Tom's Cabin* and joins forces with Penguin Classics to provide the most comprehensive library of world literature available from any paperback publisher.

1990s The launch of Penguin Audiobooks brings the classics to a listening audience for the first time, and in 1999 the worldwide launch of the Penguin Classics Web site extends their reach to the global online community.

The 21st Century Penguin Classics are completely redesigned for the first time in nearly twenty years. This world-famous series now consists of more than 1,300 titles, making the widest range of the best books ever written available to millions—and constantly redefining what makes a "classic."

The Odyssey continues . . .

The best books ever written

PENGUIN CLASSICS

SINCE 1946

Find out more at www.penguinclassics.com

Visit www.vpbookclub.com

CLICK ON A CLASSIC
www.penguinclassics.com

The world's greatest literature at your fingertips

Constantly updated information on more than a thousand titles,
from Icelandic sagas to ancient Indian epics, Russian drama to
Italian romance, American greats to African masterpieces

•

The latest news on recent additions to the list, updated
editions, and specially commissioned translations

•

Original essays by leading writers

•

A wealth of background material, including biographies
of every classic author from Aristotle to Zamyatin, plot
synopses, readers' and teachers' guides, useful Web links

•

Online desk and examination copy assistance for academics

•

Trivia quizzes, competitions, giveaways, news on
forthcoming screen adaptations